Maria Boştenaru Dan, Alex Dill (Eds.)

Water as hazard and water as heritage

Report of the European Geosciences Union Topical Event
in Rome, 13.-14. June 2016

Water as hazard and water as heritage

Report of the European Geosciences Union Topical Event in Rome, 13.-14. June 2016

Edited by
Maria Boştenaru Dan and Alex Dill

KIT Scientific Publishing

Photos front cover by Maria Boştenaru:
right: Flood mitigation in Bernburg, Germany, 2013
left: Water parterre at villa Gamberaia, Florence,
designed by Romanian princess Ioana Ghyka

Printed with the support of the Marie Curie Alumni Association.

Impressum

KIT Scientific Publishing

Karlsruher Institut für Technologie (KIT)
KIT Scientific Publishing
Straße am Forum 2
D-76131 Karlsruhe

KIT Scientific Publishing is a registered trademark
of Karlsruhe Institute of Technology.
Reprint using the book cover is not allowed.

www.ksp.kit.edu

Print on Demand 2018 – Gedruckt auf FSC-zertifiziertem Papier

ISBN 978-3-7315-0736-9
DOI 10.5445/KSP/1000076621

Acknowledgements

We would like to thank the European Geosciences Union for their support (S2015/255) of an independent workshop following the series of sessions on "Natural hazards' impact on urban areas and infrastructure" which is still running at the EGU General Assembly and where we presented the results.

We acknowledge the support of:
http://www.egu.eu/meetings/co-sponsored-meetings/
http://www.egu.eu/meetings/support-request/

We also thank the Accademia di Romania a Roma for providing the spaces for the Workshop, Prof. Dr. Mihai Bărbulescu, during the Vasile Pârvan scholarship (supervisor prof. dr. arch. Cristina Olga GOCIMAN) at this institution where the workshop took place. The book publication was made possible with the support of the Marie Curie Alumni Association. Work towards the book also benefitted from short visits in Karlsruhe throughout 2016/17 with the support of the COST action TD1406 "Intelligent management of heritage buildings" and the Marie Curie Alumni Association and in 2017/18 through the Romanian funding scheme Mobility of Researchers grant number PN-III-P1-1.1-MC2017-1334. We acknowledge also the participation of Marie Curie alumni in the Workshop (Maria Bostenaru Dan, the organizer, Mirela Adriana Anghelache, Mario Montopoli and Michele Norri).

Dr. Dipl.-Ing. Maria Bostenaru Dan

Forscherin an der „Ion Mincu" Universität für Architektur und Städtebau
und internationale Forscherin des KIT, Gastwissenschaftlerin

Akademischer Oberrat Dipl.-Ing. Alex Dill

Leiter des Fachgebiets Architektur und Mobiliar des KIT

Theme

In a changing climate, hydrological and meteorological hazards related to water provoke more and more losses. Water courses are also causing other types of hazards, as alluvial soil deposits are raising vulnerability to earthquakes through the Mexico City effect. On the other hand, water itself is a vulnerable habitat. To deal with the latter, living museums including aquaria are planned, to raise awareness of the need to protect the ecological diversity of water and water sites. Protecting water sites can also be, at times, protection against water related hazards, as landscape architecture began to recognize recently. Floodplains are an alternative to dams and dikes. This would be already enough to emphasize water as a dual element, but the duality goes further. Having to think about duality, we think of Chinese philosophy. Water (or the lack of it, as in drought) destroys life as a hazard, but gives life as well. To this symbol of giving life is connected the role of water as heritage, to the development of culture next to water. Architecture of river and coastalscapes underscore this. Leisure architecture is connected to this, and includes also architectural objects such as baths.

Recently there were a couple of initiatives dealing with subjects other than the natural hazards view of natural sciences and engineering:

- the Water Junior Summit of the European Science Foundation looking to water from a multidisciplinary point of view in Stresa, Italy, in 2012
- The Water symposium in Japan for women scientists at Hokkaido University, since 2014, in a 5 year program
- The seminar by Alina Payne from Harvard held in the extended Mediterranean space "From Riverbed to Seashore" in 2013-2015
- The workshop on "Floods, dams and dikes in early modern world" at New Europe College, Romania in 2014.
- Water forms: History, Gestalt and Semantics of an Element, at Kunsthistorisches Institut Florenz, in 2016.

Approaches to hydrological and meteorological hazards within the EGU are numerous. The aim of this Workshop was to bring together selected participants from these events to examine the duality, thus far neglected in EGU events. The organizer was conducting research for a scholarship on the topic of adaptation of architecture programs to the needs of women pioneers in architecture as investor, planer/designer and user, in the coming academic year. The form of the workshop was the science café; after presentations of the key note speakers the participants will gather at a round table and brainstorm on the a set of preselected questions.

Dr. Dipl.-Ing. Maria Bostenaru Dan

Forscherin an der „Ion Mincu" Universität für Architektur und Städtebau
und internationale Forscherin des KIT, Gastwissenschaftlerin

Akademischer Oberrat Dipl.-Ing. Alex Dill

Leiter des Fachgebiets Architektur und Mobiliar des KIT

Contents

Workshop organisation

Maria Bostenaru Dan[1]*, Alex Dill*[2]

[1] *Vasile Pârvan alumna, Accademia di Romania a Roma, Italien*
[2] *Karlsruhe Institute for Technology, Karsrluhe, Germany*

Venue

Accademia di Romania a Roma
Piazza Jose de San Martin 1
00197 Rome
ITALY
http://www.accadromania.it/
https://sites.google.com/site/egutopicalevent/

Contact

Organizer of the workshop (awardee of the EGU topical event)
Dr. Dipl.-Ing. Maria BOSTENARU-DAN
Vasile Pârvan fellow at Accademia di Romania a Roma
email: Maria.Bostenaru-Dan@alumni.uni-karlsruhe.de
phone: 0040 771 225139 (Romanian number)

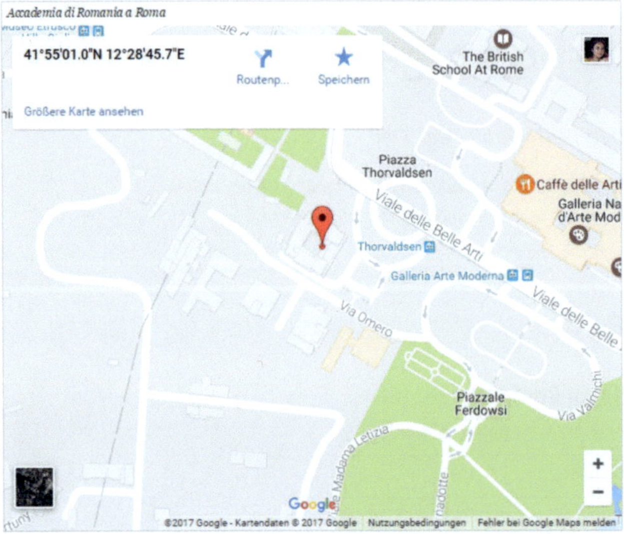
Accademia di Romania a Roma
41°55'01.0"N 12°28'45.7"E
Routenp...
Speichern
Größere Karte ansehen
The British School At Rome
Piazza Thorvaldsen
Caffè delle Arti
Viale delle Belle Arti
Thorvaldsen
Galleria Arte Moderna
Via Omero
Piazzale Ferdowsi
Via Valmichi
©2017 Google - Kartendaten © 2017 Google Nutzungsbedingungen Fehler bei Google Maps melden

ACCADEMIA DI ROMANIA

Committees

Organising committee

Dr. Maria Bostenaru-Dan (RO-IT) – convener
Dr. Roxana Mihaly (RO-IT) – posters
Dr. Eva Pietroni (IT) - guided visit
Ciprian Buzilă (USA) - guided visits

Programme committee

Dr. Maria Bostenaru Dan (RO-IT)
Prof. Dr. Cristina Olga Gociman (RO)
Oberakademischer Rat Alex Dill (DE)
Assoc. Prof. Dr. Cerasella Crăciun (RO)

Participants

1. Dr. Marie Beatrice Andreucci is adjunct professor at the Faculty of Architecture at Sapienza University in Rome, in environmental design.
2. Mirela Anghelache, geologist, is researcher at the Romanian Academy and Marie Curie alumna.
3. Dr. Gruia Bădescu, geographer, is lecturer at the University of Oxford. He obtained his doctorate at the University of Cambridge, UK.
4. Dr. Beatrice Brizuela is civil engineer and researcher at INGV (Istituto Nazionale Geofisica e Volcanologia), Rome. She holds a doctorate in solid earth physics from the University of Bologna.
5. Ciprian Buzilă, doctoral candidate at Brown university, USA, is Rome scholar alumnus and expert in the historical studies of monuments. He holds a master and two bachelor degrees in architecture and arts.

6. Dr. Daniela Calciu is an architect and assistant professor at the "Ion Mincu" University of Architecture and Urbanism, where she obtained her doctorate.
7. Dr. Oana Diaconescu is an architect and lecturer at the "Ion Mincu" university, with a joint doctorate from this university and the University of Genoa in Italy. She holds a master in Architecture and Archaeology from Rome.
8. Mark Andrew Kelly is an architect and fellow of the British School in Rome in 2015. He obtained a bachelor's degree in Glasgow and a master's in Berkeley, USA.
9. Dr. Marina Mihăilă is an architect and lecturer at the "Ion Mincu" University of Architecture and Urbanism, where she obtained her doctorate.
10. Dr. Roxana Mihaly is a graduate in European relations and photography and former Vasile Pârvan Fellow at the Accademia di Romania a Roma, earning her doctorate from Sapienza University of Rome.
11. Dr. Mario Montopoli is a researcher at CNR-ISAC (National Research Council of Italy-Institute of Atmospheric Sciences and Climate) in Rome. He holds a doctorate in technologies for remote sensing from the University of Basilicata, Italy. He was a former Marie Curie fellow at the University of Cambridge (UK) in remote sensing of volcanic clouds.
12. Dr. Beniamino Murgante is professor for of urban planning at the University of Basilicata. He obtained his doctorate from the University of Pisa.
13. Dr. Michele Nori is Marie Curie Fellow at the European University Institute, Florence. He obtained a doctorate in rural sociology at the University of Wageningen, the Netherlands.
14. Dr. Patricia Osmond, historian, is Affiliate Assoc. Professor, Iowa State University, and former Resident Director of the ISU Rome Program. She is also Curator of the Archive and Fototeca at the Villa Gamberaia, Florence.
15. Roxana Triboi is a landscape architect and doctoral candidate at "Ion Mincu" University.

16. Dr. Maria Bostenaru Dan is an architectural engineer from the University of Karlsruhe, Germany, Fellow of the Romanian Academy in Rome and Marie Curie alumna; she received her doctorate from "Ion Mincu" University.

Questions discussed

1. Mapping and assessment methods of the impact of water related hazards on urban areas (prevention and post-disaster)
2. Techniques to visualize and communicate the evaluated impact, including GIS, internet, 3D.
3. Strategies to reduce the impact of water related hazards on urban areas.
4. Suitable methods of urban design to mitigate the effects of disasters in multi-hazard cases.
5. Partnership models between the actors involved in the decision process to mitigate disasters.
6. Urban planning instruments for strategies of risk management (ex. master plans)
7. Lessons learned about the relationship between hazard, vulnerability and impact on recent events.
8. Investigation of urban morphology for a better estimate of urban vulnerability (building site interactions)
9. Investigation of urban morphology to evaluate the post-disaster accessibility of strategic buildings, the role of street patterns for emergency vehicles.
10. Models to quantify vulnerability through questionnaires - the role of statistics.
11. Interactions between urban sub-systems which may increase/diminish vulnerability.
12. The difference between approaching the impact in protected urban areas as compared to normal areas.

13. Preserving the memory in the reconstruction/reconfiguration efforts after disasters, the role of heritage habitat.
14. Water as a vulnerable habitat for different life forms.
15. Flood protection through landscape planning interventions.
16. Bringing cities to the river or far from the river.
17. The role of water in gardens - from beauty to ecological questions.
18. Water in architecture - the relationship of the built shape to water, for example in thermal baths.
19. Water as source of renewable energy.
20. How water is represented in arts, and especially in newer media such as photography and digital media (see also excursion). What do phototeques offer on past hazard events and heritage approaches?
21. Oriental and European approaches to the semantics of the element (see excursion).

Programme

Day 1 - 13 June 2016

9:00-9:30	Administrative matters
9:30-13:00	Presentations (20 min each). Moderator: Maria Bostenaru
	Maria Bostenaru (Accademia di Romania a Roma): Introduction to the topic
	Anca Cezarina Fulger (Accademia di Romania a Roma): Water in the Roman Emporium: typology and structure
	Oana Diaconescu ("Ion Mincu" University of Architecture and Urbanism, Bucharest): Water and Roman Archaeology

Daniela Calciu ("Ion Mincu" University of Architecture and Urbanism, Bucharest): Waterways as carriers of art and repositories of heritage

Patricia Osmond (Iowa State University): Villa Gamberaia and the *parterre d'eau* of Princess Ghyka (1896-1925): water as heritage in a potentially hazardous environment

11:30 - 11:45 Coffee break

M. Beatrice Andreucci (Sapienza University of Rome): Heritage and Water. Symbols and Challenges in the works of Carlo Scarpa.

Gruia Bădescu (University of Oxford): Back to the Waterfront: Urban Post-War Reconstruction and Waterside Development in Belgrade and Beirut

Marina Mihăilă, Cristian Bănică ("Ion Mincu" University of Architecture and Urbanism and Centre for Studies of Contemporary Architecture, Bucharest): Migration in waterscapes: from royal model to new-scapes in Copenhagen (Abstract at the end of the page)

Mario Montopoli (National Research Council of Italy-Institute of Atmospheric Sciences and Climate): A weather radar approach for rain precipitation estimation in urban areas

Beniamino Murgante (University of Basilicata, Potenza): Smartness, technologies and natural risks.

13:00-15:00	Break
	Guide: Roxana Mihaly
	Posters (Marina Mihăilă, Roxana Triboi, Oana Diaconescu, Mirela Adriana Anghelache, Mark Kelly - British School of Rome)
	Visit of Spazi Aperti exhibition at the Accademia di Romania a Roma
15:00-16:00	Discussion - Science café. Moderation: Ciprian Buzilă
	Mirela Anghelache (Romanian Academy), Anca Cezarina Fulger (Accademia di Romania a Roma), Roxana Triboi ("Ion Mincu" University of Architecture and Urbanism), Ciprian Buzilă (Brown University).
16:00-17:00	Book launch of "Space and time visualisation" (editors Maria Bostenaru Dan and Cerasella Crăciun), Springer International, Switzerland, 2016.
17:00-18:00	Presentation of the visit to Villa Giulia. "Virtual Museum of Tiber valley" project presented by Eva Pietroni (National Research Council of Italy - Institute for technologies applied to cultural heritage) (Abstract at the end of the page)

Day 2 - 14 June 2016

9:30-13:00 Museum visits for water related approaches

- Spazi aperti exhibition of Accademia di Romania a Roma
- Villa Giulia
- Villa Borghese with the other academies
- Terme di Diocleziano / Diocletian's Baths with water exhibition

Impressions

Figure 1: Group photo by Roxana Mihaly.

Figure 2: Introduction of the guided tour by Eva Pietroni. Photo: M. Bostenaru

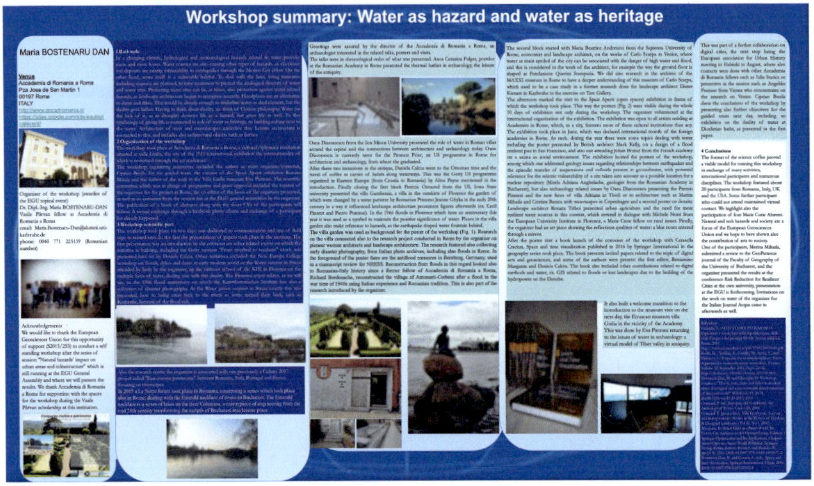

Figure 3: Poster at the EGU GA 2017

Media

[1] Report on the event on the Le Notre website http://ln-institute.org/news/news_show_details.php?news_id=642

[2] Presentation at the RRRC (Risk Reduction for Resilient Cities) conference, Bucharest, 4-5 November 2016, with publication

[3] Review by Marina Mihăilă in Geopatterns journal http://www.geodinamic.ro/assets/geo-patterns/volumes/v1.2_42-43.pdf

[4] Presentation at the EGU GA 2017 http://meetingorganizer.copernicus.org/EGU2017/EGU2017-191-3.pdf

[5] Presentation of the book at the EGU GA 2017 http://meetingorganizer.copernicus.org/EGU2017/session/26228

Spazi aperti

Roxana Mihaly

Curator of the Spazi Aperti exhibition, Accademia di Romania in Roma

Extended abstract

Spazi Aperti (Open spaces) is a project started in 2003 by the "Vasile Pârvan" fellows at Accademia di Romania in Roma with the aim of creating a dialogue with the artists in residence at the academies and cultural institutes in Rome.

Accademia di Romania in Roma is located in the centre of Rome, conducting its activity on the basis of the Cultural Agreement between the Romanian government and the Italian government. The administration of the Accademia is in the hands of the Romanian Ministry for External Affairs, its scientific activities being coordinated by the Romanian Academy and the Ministry of Education in Bucharest, and the cultural ones by the Romanian Cultural Institute. The mission of the Accademia di Romania is to make known Romanian culture outside the country and to assure its visibility and prestige. In the framework of the institution annually international conferences are organized in different fields (archaeology, history, arts, architecture etc.), book launches or events with an important impact on the Romanian community in Italy.

Spazi Aperti is the only event in Rome which aims to promote of young resident artists from the foreign academies and cultural institutes in Rome, being organized and hosted by the Accademia di Romania in Roma. From 2003 on it managed to reunite internationally renowned artists who were in Rome on visiting fellowships. Among these we mention Matthias Weischer (2007), Carsten Nicolai aka Alva Noto (2007), Emmanuel Giraud (2009),

Daniel Widrig (2009) Ana Rewakowicz (2012), Klaus Weber (2013), and Till Roeskens (2013) as well as the young artists representing Romania such as Marius Purice (2009), Ada Milea (2009), Alexandru Balanescu (2009), Sergiu Toma (2012), Robert Fekete (2012), Claudia Mandi (2013), Marius Burhan (2013), and the list could continue.

Started by the artist fellows of the academy, Spazi Aperti received numerous positive reviews, establishing itself later as one of the most visible events of contemporary art. The event gained renown particularly through the involvement of the other academies and cultural institutes present in the eternal city (The British School at Rome, Accademia di Francia Villa Medici, Accademia Tedesca Villa Massimo, The American Academy in Rome, Real Academia di Spagna, Knir, Rhode Island School of Design, Forum Austriaco di Cultura Roma, Accademia di Danimarca, Istituto Svizzero, Accademia d'Ungheria in Roma - Istituto Balassi, Accademia Belgica Roma, etc.). This particularity enhanced the visibility and importance of the Accademia di Romania in Rome in the Italian cultural milieu. In the course of time the event was included in contemporary circuits such as Roma Contemporanea. In 2012 one of the main conferences of the event took place at Palazzo delle Esposizioni.

Each year the Spazi Aperti event attracts a young public of over 1000 persons Romanians as well as Italians, Spanish, Germans, English, and Americans etc. It has also attracted prominent persons from the world of contemporary art, including artists, contemporary art collectors, artistic directors, cultural journalists etc.

Spazi Aperti was intended from the beginning as a representative event for both the Accademia di Romania in Roma, and the image of the Romanian state in the international cultural and artistic context. The former editions were praised for the high-quality contribution to the Romanian cultural milieu through the presence of leading artists, as well as the high-level presentation and organization of the events.

Why Spazi Aperti? Because the event does not take place just in the exhibition hall or in the conference hall but also in the courtyard, garden and corridors. All spaces of the building, including the non-conventional environment of the Accademia di Romania, are at the disposal of the artists.

The XIVth edition of Spazi Aperti, which took place between 10-19 June 2016, tried to include as part of the international exhibition of contemporary art a series of original cultural events. The vernisage, which took place on the 10th of June 2016, was realized with a Flash MOD dedicated to the support of young talent as well as the performing arts. The Flash Mob and the concert were performed by Orchestra Giovanile di Monte Mario (O.G.M.M.), under the supervision of M Alfredo Santoloci (artistic director of the orchestra and at the time the director of the Music Conservatory "Santa Cecilia" in Rome). The evening ended with a live performance of contemporary music.

The second event in the framework of the XIVth edition of Spazi Aperti was dedicated to the Workshop "Water as hazard and water as heritage", event organized by the "Vasile Pârvan" fellow Maria Boștenaru–Dan. It included speakers from different disciplines (architecture, history of art, landscape, geography, geology, meteorology, urban planning, etc.). After the talks the poster section in the exhibition dedicated to the Workshop was visited. Finally there was a concluding round table. Also, in the context of the Workshop the book launch "Space and time visualisation" (Springer) took place.

The third evening was dedicated to theatre, when the multidisciplinary action "Noli me tangere" took place, conceived by Wanda Mihuleac and performed by the actor Davide Napoli, with music by Cornelia Petroiu, and dance by Isabelle Maurel.

The fourth evening was dedicated to Jazz, so that the day started with a conference by prof. Virgil Mihaiu (Universitatea de Muzică "Gh. Dima" din Cluj) together with Francesco Martineli: "Prospettive italo-romene nel ambito del jazz europeo / Dialogo interculturale fra i jazzologi". The day dedicated to Jazz closed with a concert of the Italian orchestra Bixi Big-Band together with the American singer Tess Amodeo-Vicker.

Spazi Aperti 2016 XIVth edition closed on the 19th of June with an accordion concert performed by the young musician, student of the Conservatory " S. Cecilia" in Rome, Samuele Telari.

The topic of the XIVth edition was "The code of fortune", the artists having the freedom to express their vision about the evolution and laws of life. Along the course of centuries, artists have given to humanity millions of symbols, which over time have become real talismans of fortune or carriers of messages. Meetings, influences, coincidences, choices, errors in collective thought – all these are hidden in the genetic fingerprint of each of us or in the influences of a code of fortune. Simultaneously, during their stay in Rome, the fellows of different academies were influenced by the concept of "open air museum", by the beauty of Rome which hides so many secrets.

The events organized in the interior of the Accademia di Romania in Roma hosted the works of about 50 artists – fellows of the academies and foreign institutes in the Eternal City. At the same time, the XVIth edition also had a section of special guests, namely the Italian maestro, considered the father of informal art, Achile Pace, and from Romania, Mihail Trifan.

Vita

Roxana Mihaly was born in 1988 in Tîrgu Mureș. In 2010 she graduated from the faculty of International Relations and European Studies at the University Petru Maior of Tg. Mureș. During 2010-2012 she pursued a Masters program "Elites, Culture and European Construction" at the same university. In the last year of the program she benefited from an Erasmus scholarship at the "La Sapienza" University in Rome, and in 2012-2013 the "Vasile Pârvan" Fellowship at the Accademia di Romania in Rome. At the end of 2015 she obtained a doctorate at the "La Sapienza" University in Rome with the thesis: “The identity construction of cultural elite: Accademia di Romania (1922-1948)”.

With a passion for photography, in 2012 she graduated from the Popular School of arts of Târgu Mureș, photo-video department. She has participated in numerous collective exhibitions in Romania and Italy. In 2016 she was curator of the XIV edition of the international edition of the contemporary art exhibition “Spazi Aperti” organized by Accademia di Romania.

EGU Topical Event Summary: Workshop "Water As Hazard And Water As Heritage"

Maria Bostenaru Dan

"Ion Mincu" University of Architecture and Urban Planning, Bucharest

Abstract

On 13-14 June 2016 a Workshop to on the topic "Water as hazard and water as heritage" took place at the Accademia di Romania a Rome, with funding from the European Geosciences Union. The aim was to reunite geoscientists and artists dealing with the dual character of water. This paper presents a summary of the rationale, the organization and the findings during the Workshop for a broader audience.

Purpose

In a changing climate, hydrological and meteorological hazards related to water provoke more and more losses. Water courses are also causing other types of hazards, as alluvial soil deposits are raising vulnerability to earthquakes through the Mexico City effect. On the other hand, water itself is a vulnerable habitat. To deal with the later, living museums including aquaria are planned, to raise awareness to protect the ecological diversity of water and water sites. Protecting water sites can be, at times, also protection against water-related hazards, as landscape architecture has begun to recognize recently. Floodplains are an alternative to dams and dikes. This would already be enough to underline water as dual element, but the duality goes further. Having to think about duality, we think of Chinese philosophy. Water (or the lack of it, as in drought) destroys life as a hazard, but gives life as well. To this symbolism of giving life is connected the role of water as heritage, to building culture next to the water. Architecture of river and coastalscapes underline this. Leisure architecture is connected to this, and also includes architectural objects such as baths.

Organisation of the workshop

The Workshop took place at Accademia di Romania a Roma, a cultural and diplomatic institution situated at Valle Giulia, the site of the 1911 international exhibition whose international spirit is continued through the academies of arts and humanities.

The Workshop organizing committee included the author as main organizer/convener, Ciprian Buzilă for the guided tours, the curator of the Spazi Aperti exhibition Roxana Mihaly and the author of the work in the Villa Giulia museum Eva Pietroni. The scientific committee which was in charge of program and grant approval included Cristina Olga Gociman the organizer's mentor for the project in Rome, Alex Dill, co-editor with the organizer of the book presented, as well as co-convener from the session run at the EGU general assemblies. The publication of a book of abstracts along with the short profiles of the participants is done with this book. A virtual exchange through a facebook photo album and exchange of a participant list already took place.

Workshop: a summary of the proceedings

The Workshop took place on two days, one dedicated to communication and one to field trips to related sites. On the first day presentations of papers took place in the morning. The first presentation was an introduction by the convener on other related events on which the initiative is building, including the Getty seminar "From riverbed to seashore" which was presented later on by Daniela Calciu. Other initiatives included the New Europe College workshop on floods, dykes and dams in the early modern world or the Water summit in Stresa attended by both the organizer and other Marie Curie alumni, up at the summer school of the KHI in Florence on the multiple facets of water, dealing also with this duality. The Florence event relates, as we will see, to the anniversary of the 1966 flood on which the Kunsthistorisches Institute also has a collection of disaster photography. At the Water junior summit in Stresa (Figure 4) exactly this issue was presented, how to bring cities back to the rivers as some turned their back, such as Karlsruhe, because of the flood risk.

The research centre with which the organizer is associated also ran previously a Culture 2007 project called "Eau comme patrimoine" between Romania, Italy, Portugal (Figure 5) and France focusing on riverscapes.

In 2015 a Le Nôtre forum took place in Romania, continuing a series which also took place in Rome, dealing with the Emerald necklace of rivers in Bucharest. The Emerald necklace is a series of lakes on the river Colentina, a masterpiece of engineering from the mid-20th century transforming the swamps of Bucharest into leisure place (Figure 6).

Greetings were offered by the director of the Accademia di Romania a Roma, Mihai Bărbulescu, an archaeologist interested in the related talks, posters and visits.

Figure 4: Stresa, Lago Maggiore. Photo: M. Bostenaru.

Figure 5: Coimbra. Photo: M. Bostenaru.

Figure 6: Colentina lake of Herăstrău. Photo: M. Bostenaru.

First block

The talks were in chronological order of the topics presented. Anca Cezarina Fulger, postdoc at the Romanian Academy in Rome presented the thermal baths in archaeology, the leisure spaces of antiquity (Figure 7).

Oana Diaconescu from the Ion Mincu University presented the role of water in Roman villas around the capital (Figure 8) and the connections between architecture and archaeology today. Oana is currently tutor for the Piranesi Prize, a US program in Rome for architecture and archaeology, from which she graduated.

Figure 7: Therme di Diocletian. Photo: M. Bostenaru.

Figure 8: Therme di Diocletian. Photo: M. Bostenaru.

After these two incursions into the antique, Daniela Calciu went on to the Ottoman period and the transport of coffee as a carrier of habits along waterways. This was the US Getty program organized in Eastern Europe (from Croatia to Romania) by Alina Payne mentioned in the introduction.

Finally, closing the first section, Patricia Osmond from Iowa State University presented the Villa Gamberaia, a villa on the hills overlooking Florence whose old-fashioned *parterre* was transformed into a magnificent new *parterre d'eau* (Figure 9) by the Romanian Princess Jeanne Ghyka in the early 20th century in a way that later influenced prominent landscape architects in the United States and Europe. After the devastating 1966 floods in Florence, which had an anniversary that year, the architectural photographer Balthazar Korab contrasted the raging waters of the Arno with the quiet pools of the Gamberaia, recording in his photo essay the positive significance of water. Places in the gardens also make reference to hazards, such as the figure of Poseidon as god of underground forces and earthquakes and the deliberately inclined pilasters simulating the seismic effect of an earthquake, at the entrance to the grotto.

The villa garden was used as background for the poster of the workshop (Figure 10). Research on the villa was also connected to the research project conducted in Rome by the organizer on pioneering women architects and landscape architecture. The research also featured collecting early disaster photography, from Italian photo archives, including floods in Rome. In the foreground of the poster there are the anti-flood measures in Bernburg, Germany, used in a manuscript review for NHESS. Reconstruction following floods in this regard also looked to Romanian-Italian history since a former fellow of Accademia di Romania a Roma, Richard Bordenache, reconstructed the village of Antonești-Corbeni after a flood in the war years of the 1940s using Italian experience and Romanian tradition. This is also part of the research introduced by the organizer.

Figure 9: Villa Gamberaia, Settignano, Florence.

Figure 10: Poster of the workshop

The second session started with Maria Beatrice Andreucci from the Sapienza University of Rome, economist and landscape architect, on the works of Carlo Scarpa in Venice (Figure 11), where water as the main symbol of the city can be associated with the danger of high water and flood, and thus considered in the work of the architect, for example the way the ground floor is shaped at Fondazione Querini Stamparia. We also did research in the archives of the MAXXI museum in Rome to have a deeper understanding of this museum of Carlo Scarpa, which used to be a case study in a former research project done for landscape architect Dieter Kienast in Karlsruhe in the exercise on Tate Gallery.

Museum places – Carlo Scarpa - Fondazione Querini Stampalia

A museum is an institution which serves most of the time for the exhibition of collections, but also for their conservation, research and valorization. Their goal is the joy, the education of people. The architect adapts his style to the program, which means that the architectural topic should be enlarged correspondingly so that the desired spiritual atmosphere is achieved. A museum is a pause in the run of time and space. There is a focus, which has an influence in social life, but which can be found again in the dreams of each man. The exhibited objects should recall these dreams and through this there is an escape from time and the space. The museum is through this a special space and the visit to the museum a unique experience.

Traditionally museums were formed in palaces, as their monumental architecture could mark exactly this monumentality. Modern architecture has not made use of this in most cases and it is difficult in the frame of the multitude of museums that seem to be mere containers or display windows to find some which have still a particular expression.

Figure 11: Carlo Scarpa architecture in Venice. Photo: M. Bostenaru

This is different at in the case of Scarpa. He calls the topic of the palace back and gives it a new personality. Palazzo Querini was originally built in the 16th-century Renaissance style. A badly done restoration in the 19th century took the original character away and changed, especially in the interior, this to a classicizing style The façade remained unchanged. Scarpa reorganized particularly the interior, with a focus on the entrance zone, which signifies the transition of space and time. Before this the two entrances were exactly the opposite. Towards "campiello", the campo on the side of the church Santa Maria Formosa, where the building through the channel with the same name has its boundary there was only a service entrance. The main entrance from the land was on the side of the garden, where today there is only a hidden door. Scarpa references in this transformation the motifs of Venetian bridges and the entrance is an interpretation in the sense of contextualizing old built substance.

The old Venetian bridges lead without parapet over the channels. Scarpa foresees a parapet, but the shape and the materials underline the separation of constructive elements. Particularly in case of bridges the crossing and over tensioning are clear. As entrance he uses a former window. The two deeper placed vaulted gates which always have problems of flooding remain further entrance to water, for the flood and the air. Through this the former entrance hall will become a transit zone between interior and exterior, the first meeting of Scarpa's concept with this relationship which crosses the whole design.

Many references exist to the role of water in the design of Scarpa. Water plays a role in the contextualist approach, water as Venetian motif, the source of —inspiration, water as transparent shaping element, although when the commission was given it was identified as the main problem. I would like to underline the symbolic meaning of water. Water as flood is what one leaves in order to visit the exhibition. The exhibition hall is built as a transit space – its ends are glass walls and the conformation of both Porticos displays clearly the transit zone character, no longer physically, but symbolically. The two heating units were introduced as sculptural elements, and are so accentuated exactly as memory. Dreaming during the visit is limited in time; the real life waits at the end. In the garden the motif of the flood again takes over, on a smaller scale, as the garden has in any case a different character than urban places.

And here starts a new symbolism. The garden has not to be stepped in, more does rather, it seems like a Japanese garden, one to be looked at, a further image on the exterior wall of the exhibition, the image of life. The garden is a game of architecture, of water and vegetation. While the water, despite its movement, remains the same in essence over the whole year, the vegetation changes. The relationship of the hall with the exterior introduces a new condition for adaptation – the change.

And with this we arrive at the adaptation of the program to the topic. The architecture of museums at the end of the 20th century is characterized by the spread of the program through inclusion of special functions. Here there is an extreme case, as the museum function actually is in equilibrium with the rest. On the ground floor is a hall for the exhibition of modern art and a hall for special exhibitions, on the first floor the painting exhibition. The lateral functions are represented by the library on the second floor and the conference room on the ground floor. The conference room, a room of social meeting has almost the same importance from the point of view of spatial relationship as the exhibition hall; it opens to the garden, the nodes are conceived according to function, the door, which leads there, reminds us again of Japanese architecture.

The most important extension of the museum function is the garden. This is not an additional exhibition zone, but has a symbolic function. It works as an image, as in Japanese architecture, and repeats the Venetian motif of water, the transparence of which is in accordance with the hall closed with glass. The relationship between interior and exterior occurs again over a transit space. Scarpa employs the layering as means for the spatial connection. The garden is not an urban public space but more a space for meditation and self-reflection. The museum works in this case as a filter: out of the noise of the public the exhibition space calls up dreams and sends the individual, accompanied by these dreams again into the midst of life. This is served also by the refined order of the garden: it is also a new order, which shows new leading thoughts: the order of the labyrinth, the shape, which leads to the interior, image of self-reflection.

The garden is separated from the free? room in the vicinity through a wall, separation instead of connection. To furnish the garden Scarpa uses existing elements (fountain, plants) and mostly the dominating topic of water. His composition reminds us again of the Japanese: surfaces which cannot be stepped on but are also not physically separated (ex. the grass stage), as well as the whole approach with adjacent materials. Scarpa employs layering as a means of spatial connection.

So far it was explained the way Scarpa stays against the program as own personality. Important is particularly the urban approach to the context, the Venetian motifs of water, of the bridge and of the portico; in interior the detailing of the putting next to the other the old and the new and particularly the existence of the transit space – also a motif typical for Venice (one shall think only of the numerous loggias), from a landscape point of view the return of the old oriental culture as example for what the turning towards the self of the museum spaces can mean.

Additionally for the program it shall be underlined that the façade does not betray the interior function and the plasticity of the interior spaces is in no case neutral. These build themselves a work of art and thus are not particularly suitable for the goal of the exhibition. The special exhibitions hall with the water channel can used only in extreme cases because of bad lighting.

The main reading room is lighted from both sides, the artificial lighting is realized through neon tubes which are fixed on the wall and screened through milk glass. The space episodes build a tree structure; the nodes are marked each time through levels. The rooms are integrated but regarding circulation still independent, the core is the Portico – the hall closed with glass; the garden is separated and through this private in character, in it there is the intimacy of the interior.

The typology of the exhibition space can be marked as half open – because of the transparent Portico. It is thought for temporary exhibitions of all kind, the permanent exhibition on the first floor is thought for paintings.

To the end I would like to give the reason for my choice. To me it seems important that in this project free space and constructed space are seen as of the same importance. Also important is that the core topics, as the transition room as respectively the water goes through and the heterogeneity of the details bring unity.

Second block

Gruia Bădescu, geographer from the University of Oxford looked instead into manmade hazards and the reconstruction of Belgrade and Beirut at the waterfront post war. Man-made hazards proved in 2015 and 2016 to be a major threat to heritage and academies in Rome dedicated a series of events to them, starting with the exhibition La forza delle rovine (the force of ruins) at Palazzo Altemps and continued with conferences on the near East at the British School in Rome, focusing on the near East in 2015-16 and the Balkans the year to follow, so Gruia Bădescu's contribution connects them both, and on Ruines politiques (political ruins) at the French academy, among others. It is a back to the river approach. Marina Mihăilă and Cristian Bănică presented the water related heritage in Copenhagen (Figure 12), followed by the last two presentations on hazard. First Mario Montopoli, remote sensing engineer and former Marie Curie fellow, presented the meteorology background for climate change induced precipitations. Then Beniamino Murgante, urban planner, presented the role of water in today's smart cities natural hazards approach.

The afternoon was marked by the visit to the Spazi Aperti (open spaces) exhibition in the framework of which the Workshop took place (Figure 13). This way the posters were visible during the whole 10 days of exhibition not only during the Workshop (The organizer volunteered at the international organization of the exhibition).

Figure 12: Statue of the Little Mermaid by Edvard Eriksen, 1913, in Copenhagen.
Photo: M. Bostenaru.

Figure 13: The poster session of the workshop and installation with water space and mirror of the organizer

The exhibition was open to all artists residing at academies in Rome, which, as a city, features more such cultural institutions than any. The exhibition took place in June, which was declared international month of the foreign academies in Rome. For this reason during the year there were some topics dealing with water including the poster presented by British architect Mark

Kelly, on a design of a flood-resilient pier in San Francisco, and also by Johan Brunel from the French academy on a sauna as social environment. The exhibition hosted the posters of the Workshop, among which one addressed geology issues regarding relationships between earthquakes and the episodic transfer of suspensions and colloids present in groundwater, with potential relevance not only for the seismic vulnerability of a site taken into account as a possible location for a nuclear repository (Mirela Adriana Anghelache, geologist from the Romanian Academy in Bucharest), but also archaeology-related issues (Oana Diaconescu presenting the Premio Piranesi) and the new faces of Villa Adriana in Tivoli, or architecture (Marina Mihăilă and Cristian Bănică) with waterscapes in Copenhagen and a second poster on density. Landscape architect Roxana Triboi presented urban agriculture and the need for more resilient water sources in this context, which entered into dialogue with Michele Norri from the European University Institute in Florence, a Marie Curie fellow on rural issues. Finally, the organizer had an art piece showing the reflective qualities of water: a blue room entered through a mirror.

After the poster visit a book launch (Figure 14) of the convener of the Workshop with Cerasella Crăciun, Space and time visualization published in 2016 by Springer International in the geography series took place. The book presents invited papers related to the topic of digital arts and geosciences, and some of the authors were present: the first editor (the organizer), Beniamino Murgante and Daniela Calciu. The book also included other contributions related to digital methods and water, e.g., GIS related to floods or lost landscapes due to the building of the hydropower on the Danube.

It also created a welcome transition to the introduction to the museum visit on the next day, the Etruscan museum Villa Giulia in the vicinity of the Academy. This was presented by Eva Pietroni returning to the issues of water in archaeology: a virtual model of Tiber valley in antiquity (Figure 15).

This was part of a further collaboration on digital cities, the next stop being the European association for Urban History meeting in Helsinki in August, where contacts were also made with other Accademia di Romania fellows such as Iulia Stătică or presenters in the session such as Angelika Psenner from Vienna who concentrates on research on Venice. Ciprian Buzilă drew the conclusions of the workshop and also presented further objectives for the guided tours next day, including an exhibition on the duality of water at Diocletian baths, as presented in the first paper.

Detailed information on the workshop, including some of the abstracts, is available at https://sites.google.com/site/egutopicalevent/home

Conclusions

The format of the science café proved a viable model for running this workshop with many activities, international participants and numerous disciplines. The workshop featured about 20 participants from Romania, Italy, UK and the USA. Some additional participants who could not attend maintained virtual contact. We also highlight the participation of four Marie Curie Alumni. Natural and tech hazards and society are a focus of the European Geosciences Union and we hope to have shown also the contribution of arts to society.

One of the participants, Marina Mihăilă, submitted a review to the GeoPatterns journal of the Faculty of Geography of the University of Bucharest, and the organizer presented the results at the conference Risk Reduction for Resilient Cities at her own university. A presentation was done at the next EGU General Assembly. Invitations to submit work on water from the Italian journal Acqua came in afterwards as well.

Figure 14: Space and time visualization book

Figure 15: Tiber valley installation

Vita

Dr. Dipl.-Ing. Maria Bostenaru-Dan graduated in architecture, with a focus on urban planning, at the University of Karlsruhe, Germany. Her fields of specialization concern multimedia work and project management and international cooperation, all of these in Karlsruhe as well. She spent more than a decade abroad, in Karlsruhe and in Pavia, Italy between 1996 and 2007. Funding was provided by either the German government or the European Commission (TEMPUS in Karlsruhe, Marie Curie host fellowship in Pavia, Intra-European fellowship in Pavia and reintegration grant in Bucharest), currently project evaluator for Marie Curie Fellowships along with the Romanian Ministry of Culture and ESF and Czech and French projects. Since her return to Romania 11 years ago she works as researcher at the "Ion Mincu" University of Architecture and Urbanism. Apart from this she has cooperated with international networks, such as COST actions, ESF and

ERASMUS. She spent short research stays in Portugal, Canada and Hungary. She completed a postdoctoral project from European funds at the University of Bucharest, and one on a Romanian government one for a one year stay in Rome, Italy. Research interests concern protection of heritage sites against natural disasters, especially early 20th century buildings as well as digital methods in architecture. In this sense she was admitted to a master at the FernUniversität Hagen in Germany. Volunteer experience also includes cooperation with the US (Macromedia, EERI), where she was awarded a one-month stipend at Harvard which she could not enter. Important volunteer experience regards association work, in the student dormitory in Karlsruhe, for the Marie Curie Fellows Association, as well as the completion of the reintegration grant at the Ergorom 99 Foundation.

References

[1] Farinella, R. (2013). ACQUA COME PATRIMONIO. Esperienze e savoir faire nella riqualificazione delle città d'acqua e dei paesaggi fluviali. Aracne edittrice, Roma, 2013. http://www.aracneeditrice.it/pdf/9788854859685.pdf

[2] Bridle, H. , Vrieling, A., Cardillo, M., Araya, Y., and Hinojosa, L. (2013). Preparing for an interdisciplinary future: A perspective from early-career researchers, Futures, Volume 53, P. 22–32, http://dx.doi.org/10.1016/j.futures.2013.09.003

[3] Bostenaru Dan, M. and Gheorghe, D. (2015). Workshop summary: "Floods, state, dams and dykes in modern times: Ecological and socio-economic transformations of the rural world". Web Ecol., 15, 29-31, doi:10.5194/we-15-29-2015.

[4] Osmond, P. (ed.) (2004). Revisiting the Gamberaia: An Anthology of Essays. Centro Di.

[5] Osmond, P. (guest editor) (2002). Villa Gamberaia: Sources and Interpretations. Studies in the History of Gardens & Designed Landscapes: Vol. 22, No 1.

[6] Murgante, B. (2015). Smart Cities in a Smart World. In: Future City Architecture for Optimal Living, Edition: Springer Optimization and Its Applications, Chapter: Smart Cities in a Smart World, Publisher: Springer Verlag, Berlin,, Editors: Rassia S. and Pardalos P., pp.13-35. DOI: 10.1007/978-3-319-15030-7_2

[7] Bostenaru Dan, M. and Crăciun, C. (eds.) (2016). Space and time visualisation, Springer International, Cham. DOI: 10.1007/978-3-319-24942-1

Heritage & Water: Symbols and Challenges in the Works of Carlo Scarpa

Maria Beatrice Andreucci

Sapienza University of Rome

Introduction

Certainly, water vivifies, irrigates the hearth and enables us to survive; but like any other natural element - like air, like fire, like hearth – water can also be our enemy.

It is true that Venice has always been living with water: water as an element of protection and as an element of threat; emblem of the truth of every true Pharmakon. Poison and medicine, at once.

Carlo Scarpa perfectly understood that there is no better defense than reception and shelter and – then – integration.

The symbol of the fundamental instabilitas of the sea water. But maybe of every kind of water – also the water of the rivers that "never stays". This is Carlo Scarpa's way of relating with water, caressing its dynamics and looking after its restless flow. It is not incidental that Carlo Scarpa embraced and sheltered water in all of his architectural creations. Waterways accompany Scarpa throughout his architectural path, roots of his creativity without limits.

Selected works

1952: sculptures garden – gardens of the biennale – italian pavilion venice

In 1952, Carlo Scarpa designed the Sculpture Garden for the Venice Biennale within a patio of the Italian Pavilion, originally designed in 1932 by Duilio Torres.

In his intentions, this place of passage, exposition and reflection, must present a dynamic character, related to the transit, offering at the same time well deserved rest for the visitors.

Three heavy elliptical columns support a canopy roof - shaped as if three circles would have been subtracted from a rectangle.

In his work, two apparently conflicting aspects – passage and pause - are combined in a canopy roof, designed as a sculpture, and in the underlying geometrical garden, which plays with light, shadow and water.

1954-56: venezuela pavilion – gardens of the biennale_venice

The garden patio – enclosed in the pavilion – is again pause, is a constructed garden, like any other part, and contains a basin with a gush.

For the first time, in a clear manner, all the characteristics of Scarpa's gardens take part in the project, a pleasant place, harmonically concluded, purposely created to stay, think and energize, like in a natural garden.

In particular, this garden of the Venezuela pavilion, with the bordering wall which enclose a cut of modulated light from the larger encircling garden and which makes visible the tree canopy beyond, up to the lagoon, shelter and concentrates the transparent and luminous vitality of the water in the basin.

An oasis of amenity that with the beauty of its elementary shapes, the rationality of the architecture and the musicality of the water, represents an invitation to the joy of the understanding.

1955-57: Gipsoteca Canoviana – Canova Museum – Possagno Treviso

"Undoubtedly, having a white thing – a plaster as an example – to make it stand out, one should create a dark background: it is quite spontaneous thinking like this. On the contrary, not to be polemic against the traditional rationality, but more thanks to a sudden intuition, I've observed that it would have been better creating a white background...resorting to the water to make it vibrates on the piece of the graces, giving them movement."

Carlo Scarpa

"The sun, moving on a sculpture, doesn't produce negative effects, like, instead, when in a room it goes in certain points and not in others, enlightening only certain paintings, as far as there is nothing which makes the artistic works marvelous as the powerful sun light."

Carlo Scarpa

1955-61: Veritti House – Udine

"Wright did not copy the windows from his neighbors, he introduced works of his own time; without forgetting that the essential element in Venice was and still is, water."

Carlo Scarpa

The magic of the reflecting transparencies by which the water of the external basin mirrored the house façade, entering the living room though the wide glass wall and making everything enjoyable from the windows of the bedrooms located on the first floor overlooking the living room, considering that the rooms had sliding doors, which once open created a contiguous space from the first floor through the living room and the external water basin, through the full height glass wall.

1956-74: Castelvecchio Museum – Verona

The decision to move the big fountain along the path, which leads to the main entrance, positioning it almost in front of the offices, stands clearly in all the drawings of the courtyard made by Carlo Scarpa.

Originally, the fountain was placed against the north façade, and Scarpa decides to dignify this piece of work, locating it in the center of a shallow basin, like those of the Arabian gardens. Scarpa, Venetian at heart, enjoyed very much exploiting the qualities of water.

In front of this basin Scarpa found another one and, jokingly, he put a drinking fountain in the middle, thus creating in a daily gesture of thirst-quenching, a small ceremony.

Only one person at the time can have access to the fountain, maintaining balance stepping on one foot at the time on the sole stand.

The shape of the two basins came from the observation of two large puddles on the nylon sheets laid down in the courtyard during construction.

1957-58: Olivetti Store – San Marco Square – Venice

The sculpture was realized in natural bronze and polished as a typewriter. Scarpa located the Viani's sculpture in a space which is only its own. He located it in a reflecting shallow pool of black Belgian marble, thus gaining another depth dimension.

While giving the statue an extraordinary vivifying basement – unknown in any other modern sculpture - Scarpa also located it in the most strategic visual spot of the store, most inevitably, in the convergence point of all the internal and external, lateral and vertical views.

Isolating the internal visual condition – this is the reason of the adoption of the water, apparently inexplicable – so that one is forced to go around it.

1961: Veneto Pavilion "Italia 61" – Esposizione Italia 61 – Turin

Scarpa set up the Veneto pavilion with the materials and colors of the Venetian tradition. The poetic of "sense of color and water supremacy" is interpreted integrating the two themes evocating allusive images: the material exposed is quite limited, but of outstanding quality.

The crystalline cascade of the glass chandelier – designed by Scarpa and realized by Venini – mirrors, together with a mosaic of Mario De Luigi located on a wall, on the underlying water basins, which create precious scenic effects with the play of light.

1961-63: fondazione querini stampalia – venice

Carlo Scarpa in this specific work situates all the terms – essential and marginal – of his own interpretative and poetic discourse about Venice: the water, the light, the bright and unexpected colors – the singing of the gold in the shadow – but also the whispered tones, the unmotivated loosening of an initial tension without precise destiny. This is the twitch of the stairs, landing on the bank, run out in the telescopic view of the "portego", half-height, to which the green lawn of the garden on the background is the objective sight.

Conclusive episode of this Venetian itinerary, the small secluded garden with very high walls, housing between a pomegranate and a cherry tree Cercis siliquastrum, a wellhead of the old building and a gothic lion.

Two fountains bring the sound of water in the silent quiet of this Venetian corner: the drip of the spillover on the underlying mosaic water basin. Here is where the students meet to read and talk outside. The luminous silence of the small garden, from which the slow sound of the drop is intended for the evening sparrow, sipping in the precious marble bowl.

"Inside, inside the water, like in the whole city. It's only a matter of containment, of management, of using it as a bright and reflective material; you'll see the play of light on the yellow and purple stuccos of the ceilings, magnificence!

Without forgetting that the essential element in Venice was and still is the water."

Carlo Scarpa

1966-77: entrance to the university institute of architecture iuav – tolentini santa croce venice

"The old door is laid down in a water basin, animated on the ground by steps, which run around it, forming a quoted level of a swamped landscape."

Sergio Los

1968: Monument To The Partesan – Castello Gardens Venice

The artist Murer imagined a body lying on the ground, marked by deep shadows. Scarpa taught that the best way to look at it from above would have been creating a floating iron and concrete pontoon covered with copper plates supporting the bronze sculpture so as to portray an image of a woman laying on the water surface.

1969-78: Brion Monumental Tomb – San Vito Di Altivole Cimitery – Treviso

39. The water wet stone is in the oriental art the symbol of the appearance of the mystery of life and this mystery is continuously remembered in the Brion tomb.

Not only in the evolution of the stone labyrinth, emerging from the large basin, which represents a fundamental sign of mediation among the different symbolic events, but also the continuous presence of the water as a confirmation of the centrality of the allegoric meaning, recognized to the coupling of this element with the stone.

Water flows towards the large basin, emerging from the same land where the arks have been positioned, protected by the arcosolium.

Emerging from the place of death, the water comes out circling the meditation island where the Scarpa pavilion is located, in his mind enriched by the presence of young women. The liquid element thus melts together with the images of the beginning and of the end, representing the first and the last combined.

The water is then the symbolic element of mediation among the various architectural elements. As such, it also plays the function of protecting the architecture, ritualizing the material texture.

The ground roots of the architecture are visible in the transparency, safe-guarded by the water.

The architect entrusted the water the care of architectural foundations; the building is consecrated a maternal divinity, protecting people and things alike, against the aggressions of time.

WATER AS SYMBOL OF PERENNIAL LIFE!

Vita

Registered Landscape Architect, Economist

Doctor Europaeus in Environmental Design

Adjunct Professor of Technologies for Environmental Design
and Requalification

Sapienza Università di Roma

Faculty of Architecture and Landscape Architecture

Dr. Maria Beatrice Andreucci carries out her professional activities in the fields of Green Infrastructure, landscape restoration and urban regeneration.

In parallel, her research activities are centered on the design applications of scientific and operative methods and tools for value creation in landscape transformation and requalification processes within the built environment.

Speaker at conferences, workshops and seminars, she presented papers and other works, such as projects, posters and videos in Italy and abroad (St. Petersburg, Rome, Florence, Turin, Moscow, Rio de Janeiro, Kos, Paris, Vienna, Bucharest, Tartu, Nürtingen) with interventions ranging from the evaluation of the architectural projects, landscape design and sustainability of urban transformations, to Green Infrastructure and restoration of historical gardens and parks.

She is the author of scientific and educational publications and projects of landscape architecture, with awards and prizes at international level.

Figure 16: Biennale Garden, Venice, photo credits MB Andreucci.

Figure 17: Venezia alla Partigiana, Venice, photo credits MB Andreucci.

References

[1] Giovanardi, R. (2006). Carlo Scarpa e l'Acqua, Venezia, Cicero Editore.

[2] Dal Co, F. and Mazzariol, G. (1984). Carlo Scarpa 1906-1978, Milano, Electa.

Water as Hazard – A Case Scenario of Groundwater Radionuclide Pollution in Cernavodă Area, Due to the Effects of Earthquakes

Mirela Adriana Anghelache[1], *H Mitrofan*[1], *F Chitea*[12], *and N Cadicheanu*[1]

[1] *"Sabba Ştefănescu" Institute of Geodynamics of the Romanian Academy, J.L.Calderon str. 19-21, Bucharest, mirelaadrianaa@yahoo.com*

[2] *University of Bucharest, Faculty of Geology and Geophysics, Traian Vuia str. 6, Bucharest, Romania*

Extended abstract

Relationships between earthquakes and the episodic transfer of suspensions and colloids present in groundwater are a topic with potential relevance for the seismic vulnerability of sites taken into account as possible locations for nuclear repositories. Based on a worldwide updated experience, there has been analyzed the possibility that seismic activity could significantly enhance the areal spreading of hazardous radioactive compounds which might have accidentally reached aquifers in the area of the Final Repository of Weak and Medium Active Waste (FRWMAW) at Saligny (next to Cernavodă Nuclear Power Plant, in Romania).

Figure 18: Cernavodă NPP.

The study can be considered a preliminary research in order to:

- assess the seismic risk of the zone and the possible influences on the aquifers;
- model the groundwater flow in the area and the transfer regime of radionuclides;
- assess the groundwater vulnerability in several scenarios of running the FRWMAW.

In particular, consideration has been given to the risk that seismic waves of low frequency (<0.2 Hz) would unclog pores and cracks of reservoir rocks, thereby episodically mobilizing (for time-lengths extending over several years) suspensions and colloids that are potential radionuclide carriers. For the indicated frequency range, important spectral peaks were displayed by the local micro-seismic activity. More serious detrimental effects in terms of groundwater suspensions transfer regime disturbances could be induced by certain earthquakes that were quite remote, yet strong enough for generating low frequency waves which would reach Saligny FRWMAW area.

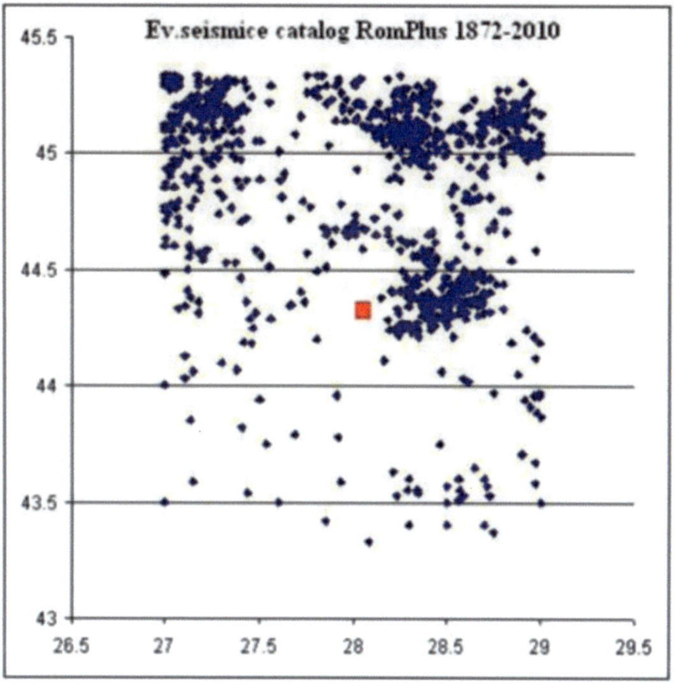

Figure 19: Cernavodă NPP.

In contrast, although many earthquakes (epicenters indicated by blue diamonds in the map above) have been recorded close to the Saligny FRW-MAW (red square on the map above) over the past 140 years, the specific spectral content of the associated waves prevented them from having a significant impact in terms of disturbances in the groundwater suspensions transfer regime.

Acknowledgements

This work was supported by the Romanian Authority for Scientific Research, project number CNMP 32-112/2008.

Vita

Mirela Adriana Anghelache completed several postgraduate schools in Architectural Conservation Studies/Built Heritage Development and Environmental Policy Curriculum (2000-2002), at Istropolitana Nova Academy, Slovakia, as well in Environmental Geology and Geophysics (1997-1998), at Bucharest University, Geology and Geophysics Department. She works as a scientific researcher in Geophysics at the Institute of Geodynamics of the Romanian Academy. Her main fields of interest are geological hazards and risks, focused on seismic hazards and risks. In this regard she held a Marie-Curie Postgraduate Fellowship in Reduction the Seismic Risk, at ROSE School, in 2004, in Pavia, Italy, and a Specialization Certificate as trained scientist in the Study and Management of Geological Risks, organized by CERG-UNIGE, in 2003, in Genève, Switzerland. She contributed as co-author to a couple of scientific articles which won the CNCSIS (National Council for Scientific Research in Higher Education) Awards in 2014 and 2010.

Back to the Waterfront: Urban Post-War Reconstruction and Waterside Development in Belgrade and Beirut

Gruia Bădescu

University of Oxford

Extended abstract

In the last decades, urban waterfronts have been the target of ambitious redevelopment projects, (re)turning cities to their river or seafronts. The paper discusses the plans, intentions and contestations of two waterside developments in two cities recovering from war and socioeconomic crisis, scrutinizing the claims that such projects are a catalyst for urban change or, in contrast, neoliberal plaster on "wounded cities". Employing a political ecology lens, the paper examines how and for whom the city is produced through socio-environmental changes and claims of sustainability and the return to water as heritage. It explores the trajectory of Beirut's seafront project, accompanying the reconstruction of its centre following the Lebanese Civil War (1975-1990), as well as the ongoing debates surrounding the Belgrade Waterfront project in Serbia's capital, described by the government as the project that would galvanize the city's urban development after the crisis of the 1990s and the bombing of the city by NATO in 1999. In Beirut, the waterfront redevelopment is based on a landfill site which emerged during the war from the debris of the ruined centre. The new district asserts to create a new urban nature on reclaimed lands. In Belgrade, a city found at the confluence of two important rivers, the waterfront project states it will finally bring the city in touch with its neglected Sava shore. Claims of sustainable architecture, a concern for ecological concepts, and the influx of capital from Gulf countries mark both projects, marketed as an urban breakthrough for the two cities and with the specific aim to reposition both as international investor destinations. Nevertheless, the projects are marked by a lack of transparency and participation, showcasing an exclusionary urbanism. Not dealing with the difficult recent past, they propose a formally generic but socially exclusive future. The political ecology lens applied to cities undergoing post-war reconstruction within neoliberal policies highlights challenges of producing urban space within discourses of sustainability but which obscure participation and transparency and prolong societal rifts.

Vita

Gruia Bădescu is a Research Associate at the School of Geography, University of Oxford. He holds a BA from Middlebury College and a MSc degree from the Cities Program at the LSE. Gruia conducted his PhD research at the Department of Architecture, University of Cambridge on urban post-war reconstruction in Belgrade and Sarajevo. He has previously worked on the uses of architecture and urban design in the process of post-war reconciliation in Beirut and Sarajevo; and on urban design and integrated urban development planning projects in Romania, Armenia, Georgia and Moldova.

Figure 20: Belgrade.

Figure 21: Beirut.

How Water is Represented in Arts, and Especially in Newer Media such as Photography and Digital Media

Ciprian Buzilă

Brown University

Extended abstract

The scope of my presentation was to respond to one of the round table questions, regarding the representation of water in arts and photography. I have chosen to illustrate water as a source of inspiration for artists and fashion designers by presenting the "In acqua: H2O molecole di creatività" exhibition which was held from May 3 to June 12 at the former Planetario of the Baths of Diocletian in Rome. The show was curated by Bonizza Giordani Aragno and Stefano Dominella.

The message of the event was a celebration of water in all of its multiple forms (liquid, gas, solid) as the true essence of life and the negative aspects of the contemporary society on this precious apparently "inexhaustible" element. The installation of Enrica Borghi suggestively entitled 'Queen of waste", dominating the center of the room and capturing the visitor's attention and made of polyethylene terephthalate bottles (PET) was a direct critique of the consumerist society and an awareness that we can recycle wastes. The message of the exhibition took form in a plethora of media, from fashion artifacts creations to installations, jewelry and photography. The Photography section, signed by photographer Paolo Belletti spoke about the journey of water flow through the Peschiera aqueduct and its ending points, the beautiful fountains of Rome and the water's capacity as life giver.

Vita

Ciprian Buzilă completed a Master's in Sacred Space and Vernacular Architecture (2011) after obtaining two Bachelor degrees: one in Art History in 2003 and one in Interior Architecture/ Furniture and Interior Design in 2010. He worked as referent at National Institute for Historic Monuments (Bucharest), scientific researcher at the Ion Mincu University of Architecture and Urbanism (UAUIM) in Bucharest and affiliate researcher at the Centre for Architecture and Urban Studies (CSAU-UAUIM). Since June 2012, he specializes in the protection and preservation of historical monuments (category A) designated by the Ministry of Culture and National Heritage and the Romanian National Institute of Heritage. He was awarded a "Vasile Pârvan" scholarship (2012-2014) and a Fulbright Junior Award (2014-2015) at University of Cincinnati.

Waterways as Carriers of Art and Repositories of Heritage

Daniela Calciu

"Ion Mincu" University of Architecture and Urban Planning, Bucharest

Abstract

The presentation will focus on the experience and research conducted within the Harvard-Getty seminar From Riverbed to Seashore: Art on the Art on the Move in Eastern Europe and the Mediterranean in the Early Modern Period. Directed by prof. Alina Payne, and financed by the Getty Foundation through the Connecting Art Histories Initiative, the seminar was a "mobile university" that cruised along the shores of the Adriatic, the Black Sea, and along and across the final reaches of the Danube and its tributary rivers, gathering six established scholars and fifteen emerging scholars from Harvard University, from Croatia, Slovenia, Poland, Serbia, Romania, Hungary, Ukraine, Turkey, and Russia. They exchanged knowledge, ideas, and experiences related to the portability of art, especially related to waterborne routs – cultural avenues that link territories despite political or military frontiers.

My research project followed the introduction of coffee to the Romanian Principalities in the last decades of the 17th century, infusing these territories with the Mediterranean sociability that became so popular and appreciated around the world.

The story of how coffee crossed over the Danube after 1659 will point to the multifaceted interpretations of the river, as religious and cultural border, as territory of connection and exchanges, especially under extreme climatic circumstances (i.e. the mini-glacial era of the 17thcentury), and as carrier and depository of heritage.

Vita

Daniela Calciu (b. 1982) is an architect and assistant professor at the "Ion Mincu" University of Architecture and Urbanism in Bucharest (IMUAM), where she obtained the PhD title based on a dissertation about the concept of "urbanity" and its role in the discourse formation of urban design after 1956. She holds a master's degree in architecture from École Nationale Supérieure d'Architecture de Paris La Villette (2007) and another in Integrated Urban Development (2009) from IMUAM, and she has been involved in applied research projects and grassroots initiatives for participatory local development, as a member of the Association for Urban Transition (since 2007). Awards and recognitions include a doctoral grant from the Romanian National University Research Council (2008-2011), the Fulbright Junior Award as visiting researcher at the University of Cincinnati (2012-2013) and the participation in a Harvard-Getty research seminar (2014-2015).

She has been collaborating with the Romanian National Museum of Contemporary Art since 2014, and she became director of development in 2016.

Water and Roman Archaeology: The Project

Oana Diaconescu

"Ion Mincu" University of Architecture and Urbanism

Abstract

The study that I proposed for the ~Spazi aperti~ will argue the need to integrate the ruins, as vital elements, which centralizes the cultural development of any city. The musealisation remains one of the few processes, compatible to the historical architecture re-use, which aims to revitalize and value the area.

The illustrative analysis will propose a series of specific terms to different disciplines, aimed to understanding and correct interpretation of the theoretical content. To this purpose, it may examine: the introduction of scientific base, the evidence of the methodology and of the indicated disciplines. There can be considered only those archaeological sites, which correspond to the need of expressing a corporate identity in relation to the type of memory generated by it. In this regard, the monument is being associated to that type of representation that can keep alive a widely recognized signified. The archaeological museum represents a particular category, because of its location on site and its connection to it, presenting as both an outdoor trail and independent entity, formally defined.

Te project will propose the representation of the relation between architectures, as an alive organism, capable of regenerating through new uses. Franco Minissi considered important for any conservation or consolidation process to consider the principles of museology and the methods of museography. The Interventions will seek to use the latest techniques and structures already studied, non-invasive and non-destructive, loose, flexible, modular and not least reversible.

For musealisation, the archaeological site will be adapted to fit to contemporary uses and to integrate new functions in relation to developments in technology integration and representation. A number of areas relating to the requirements of visitors reconfigure the historical context. The project emphasizes the relationship between public work and archaeological site exposed by detailing: input device, as a way of articulating the urban area, the service activities, the representation of the corporate identity, of teaching spaces, to site understanding, laboratories and research centers for research, libraries, ensuring knowledge, scientific approach and the permanent exhibition rooms, based on the design object. In this way, the project will emphasize to the public the new concepts of architectures related to patrimony, starting from one of the most interactive historical path through "The Fortune's code", which is the Hadrian Villa of Tibur.

Vita

Oana Diaconescu is an architect. She graduated The Faculty of Architecture at "Ion Mincu" Architecture and Urbanism University in July 2008. She holds a Ph.D. in Architecture from the UAUIM, a Ph.D. in Museography from Università degli Studi di Genova Italy (2012) and a Master's in Architecure, Museography and Archeology from the Accademia Adrianea Rome, Italy (2010). In support of her studies, she received a number of grants and scholarships among which IAESTE in Cairo, Egypt (2006) and Novi Sad, Serbia (2007), Socrates-Erasmus in Genoa, Italia (2006-2007), Ph.D. POSDRU in Rome Italy (2009-2010), Accademia Adrianea for Itinerary Master Rome (2008-2010, studying in Italy, Greece and Spain). She participated in various competitions, projects, conferences, workshops, collaborating with professionals from Romania, Italy, Spain, Portugal and USA.

Statement

Due to my continuous interest in architecture, photography and painting with I have decided to participate to the "Spazi aperti" event. I cherish the past, its traces and I think it is important to preserve and pass on to future generation our inheritance.

My curricula for the 14 years study programs as Master, Ph.D., architecture degree, etc. includes a series of exhibition an projects that showed how concept design may interact to history and patrimony.

From a deeper perspective, I became involved in the field of preservation in 2009, when I had the opportunity to win the first prize of the International Contest "Premio Piranesi- Prix de Rome", organized in Rome by the Accademia Adrianea di Architettura, Archeologia e Museografia, Roma and to receive a one-year Master scholarship in Italy. I enrolled in academic optional courses with a strong focus on monuments preservation and heritage protection. Thus, I became part of Accademia Adrianea team, being involved on projects in Romania, Italia, Spain, Greece etc. coordinating more than 150 students and wining more than 11 international prices as tutor,

I am very interested in the "Spazi aperti" event and I consider that it represents an important opportunity for my career and my research to show the humanity of contemporary architecture interventions in historical sites, their pictorial representation of synchronism and integrity.

My motivation for this taught program is also deeply related to the academic approach of the Accademia di Romania a Roma and the research opportunities on the chosen subject

As the research of the subject complement my interests and requirements, I believe that participating at this event would not only enrich my career, but would also help me to develop my maximum potentials.

Figure 22: Poster of a Piranesi workshop.

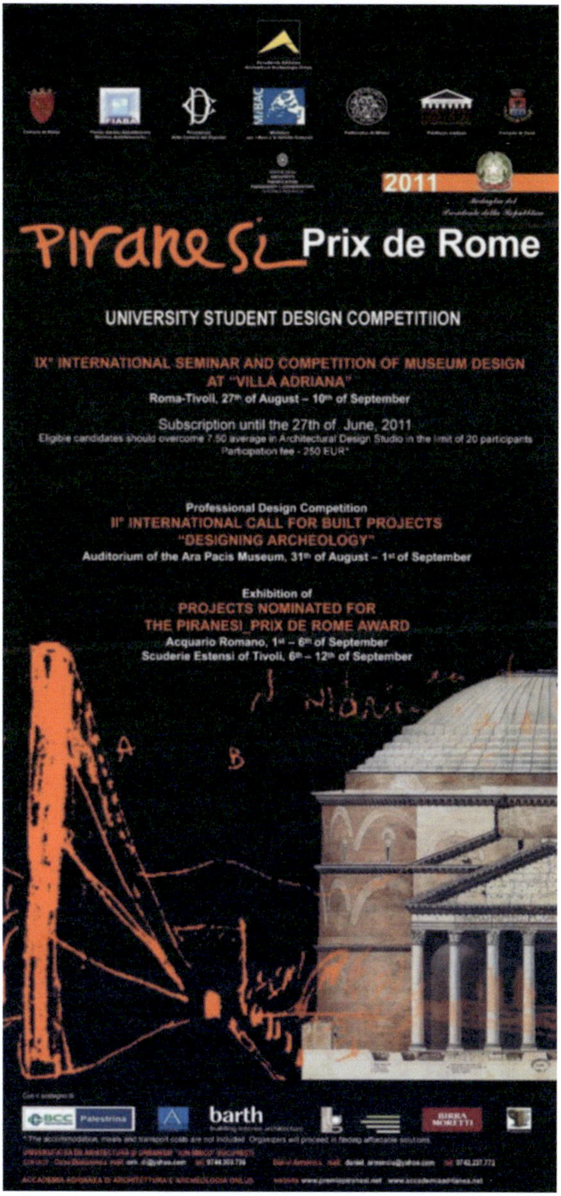

Figure 23: Poster of the Piranesi project.

Roman Baths

Anca Cezarina Fulger

Accademia di Romania a Roma

Extended abstract

Body hygiene and bathing were considered integral part of the Greek daily life being testimonied also by the Greek historical sources starting with the 8th century b.C., there where the Homerian poetry made often referrals to the man who were warm bathing in bathtubs. As a consequence of the diffusion of athletism and the bathing after physical exercises the gymnasium is developed, a place for education and social interaction in the Greek society. In the Greek society, except of the bathes in gymnasium, there were also public bathes in which an entrance fee was taken and which disposed of heating system for the rooms and the water - hypocaustum, through which the water vapors circulated slowly through terracotta tubes letting out warmth through the room floors and walls.

Depending on the habits, the physical condition and the social status the bath was a simple hygiene norm, a healthy need and of course a refined pleasure and in this sense there were different kinds of bath. Missing the rooms for bathing for body treatment lavatrinae/latrines Roman citizens could use the public bathes which were administrated by a balneator which took an entrance fee.

Outgoing from that private space lavatrina and once the heating system hypocaustum was discovered in the 1st century b.C., the Roman civilisation and the Roman architectural genius developed a program of great complexity, namely that of the thermal bathes. Consequently, the thermal bathes were built after a classical cliché which had to contain six installations, many times on a single longitudinal axis, namely apodyterium, frigidarium, tepidarium, caldarium laconium and palestra.

During the imperial period an essential change takes place in the architectural program of the thermal bathes. In this sense, the thermal bathes become monumental, with addition of new spaces for recreation and good mood, more elaborated forms appear, which contained the bathes with all sports additions, palestra, swimming pool, gardens, porticos, and other cultural annexes. The disposition of spaces with different destinations such as exedra, libraries and game rooms are present in the large imperial thermal bathes conceived in order to assure a maximal flexibility, in this way all users were free to choose the own parcour depending on intentions and preferences. So for the Romans, the thermal bathes were not only the place to bath and take care of the body, but also a place of social life and spending free time, the place where political and cultural discussions could be conducted, where it could be assisted to public conferences and theatre spectacles, art exhibitions could be seen and sculptural masterpieces, more even one could eat and drink.

The thermal bathes become more popular so that their visit was part of the daily life of the Roman citizen and in each city of the Roman empire there was at least one public bath. Mark of the Roman civis, Roman bathes will be in the 4th century, only in Rome, 856 balnea, to these being added the 11 imperial thermal bathes. Being considered the „magnificent palace of the Roman water" the Roman thermal bathes become a public institution and bring to the mentality of citizens the real body cure through the physical exercises associated to the bath, favoring the words mens sana in corpore sano (Iuvenal, Satire, 356).

Roman thermal bathes start to be abandoned starting with the 6th century a.C. once the Goths attacked the city of Rome when the apeducts of the city are destroyed, compromising in this way the water supply for the enormous quantities needed by the thermal buildings. The big buildings abandoned and unused will become in some cases cemeteries, and in other hideaway for the pilgrims and ill becoming possession of the Middle Age church institutions. Gradually they will be completely destroyed through reuse of the building materials in construction of housing.

Roman bathing traditions continued in the eastern part of the Roman empire where the process of Romanization was more intense. In the Arab medium, different of the Christian-Byzantine one, the Roman bathes will not be transformed in religious buildings or closed, but will be integrated in the palaces of the caliph, of the Arab princes, maintaining a part of their old functions, the most important being that of body hygiene. What is lost compared to the Roman world is their public role, the role in social life especially since the public bathes (hammam) had to maintain a rigorous gender separation. Also, in some former bathes constructed in the Arab environment there are some traces of post-Roman use and the thermal bath building was sometimes transformed in storage building or corn crib.

Vita

Anca Cezarina Fulger born in Bucharest graduated from the Faulty for History of the University of Bucharest, year 2003, with focus on history of the antiquity and classical archaeology, investigating in her graduation thesis Roman architecture of the spectacle and the gladiator fights. After graduating the master studies in Greek-Roman archaeology also in frame of the Faculty of History of the University of Bucharest, in 2007, she is admitted through competition at the University Uniroma 3 in frame of the Department of Historical-artistic, archaeological and conservation studies, obtaining the research doctor title in 2011 with the thesis "Imperial art between Traian and Hadrian, through the comparison of monuments in Rome with those in Dacia". She is author of various articles which regard Roman art and imperial propaganda, the monument Tropaeum Traiani and the bridge in Drobeta, she has participated at International conferences, round tables and seminars which approached large topics of the Roman Empire, namely

Roman art and architecture, but also Daco-Roman theoretic. Since 2008 she is associated member of the Centre for Comparative Studies of the Antique Societies, University of Bucharest, and in 2012 she obtained a study scholarship of the German Archaeology Institute in Rome. In 2015 she obtained through competition the Vasile Pârvan fellowship in frame of Accademia di Romania in Roma for two years, where she focuses on topics regarding Roman imperial art during Traian, organizing in frame of this institution two international conferences: Roman Imperial Propaganda, I-III centuries (21-22 October 2016) and Dacia Capta and Emperor Traian (29-31 May 2017).

Sketchbooks from Rome

Mark Kelly

ARB, RIBA, RIAS, SBA, NCARB, LEED AP RIBA Giles Worsley Rome Fellowship Prize in Architecture 2015, The British School at Rome

Abstract

"Sketchbooks from Rome" schizzi da Roma seeks inspiration from the City, the past and the present. Rome is an enduring city which has been tested over time. Drawings from the City are intended to work with the Spazi Aperti XIVa Edizione 2016 exhibition by showing the city at different stages of development. The early site sketches are intended to be exhibited next to the modern drawings, to show the forces which shaped the city, as a living place where modern Rome co-exists with ancient monuments, creating a memory of the place, a genius loci, where the fortune code and the laws of life are demonstrated. Each drawing was created onsite as close as possible to the original buildings, where you can see the individual builders hands and chisel marks in the stone, holding a special memory of those who worked to sculpt and craft this eternal city. The exhibition is composed of sketches and a sketchbook from the city, highlighting particularly interesting areas of craftsmanship.

Water and coastal cities are inextricably linked. In this San Francisco based project the design presents a vision for a city typology which can cope with sea level rise through floating water-based building structures. The inflatable foundations of these structures allow the viewing platform to rise and fall with the sea, allowing people to co-exist with environmental sea-level changes. The project maps predicted sea level rises onto a modern coastal city and presents an alternative living arrangement to co-exist with rising watermarks.

Vita

Mark Andrew Kelly is a Registered Architect currently based in central London with 9 years professional experience. I was born in Belfast, Northern Ireland. I studied at University of California, Berkeley as an Ambassadorial scholar, Newcastle University and the Mackintosh school of Architecture in Glasgow. I have taught Architectural classes in hand-drawing at UC Berkeley and have worked in Newcastle, London, San Francisco, Berkeley and Leh, India. As the 2015 BSR RIBA Giles Worsley Rome Fellow in Architecture at the British School at Rome, I researched the Republican and Imperial Roman development of concrete vaulting construction techniques and how these construction methods can inform modern use of concrete in contemporary architectural practice. The BSR exhibition builds on Mark's previous material-based research into natural materials, where at UC Berkeley Mark developed a material and method to 3D print using salt which is 96% faster than the nearest commercial alternative, faster and clean. Upon completion of his thesis, Mark's 3D printing method changed the way 3D prints are made at UC Berkeley's Architecture department, which is still being used today by students and staff as an improved process. Mark was invited back to Berkeley as a Lecturer in Architecture, leading undergraduate architectural classes for 3 years. The BSR research focuses on unreinforced concrete, using pozzolana-based concrete which was developed around Naples around the late 1st century BC and used to create the oldest remaining free-standing concrete dome in the world, the Temple of Mercury in Baiae in the Bay of Naples. Without the knowledge from the Temple of Mercury (late 1st century BC), some 200 years before the pantheon, Agrippa would have failed to build the Pantheon's coffered dome. The project focuses on the specific construction means and methods which allowed the empire to develop, construct underwater vaulted bridges on rammed-concrete-piles to conquer foreign lands and expand the empire. Barrel-vaulting, cross-vaulting and domes were construction forms designed for pozzolana-based concrete, working with the compressive strength and relative tensile weaknesses to create free span unreinforced enclosures. The BSR exhibition explores concrete construction in Roman barrel vaults, cross-vaults and modern lightweight domes. Mark's 2015 December Mostra exhibition presented works made in various mediums including drawings, blueprints, graphite sketching, watercolour paintings, cast scaled models, 3D printing, oil painting and measured drawings on fine drafting film.

Figure 24: British School in Rome

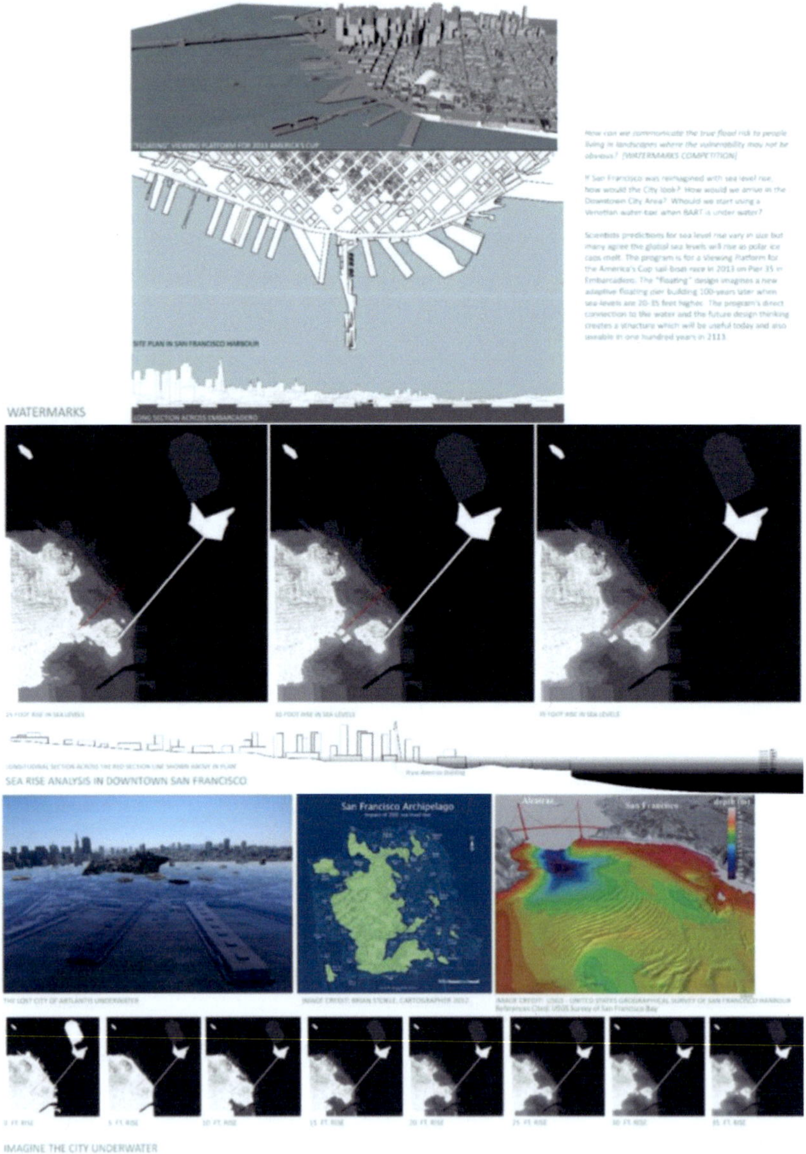

Figure 25: Watermarks part I.

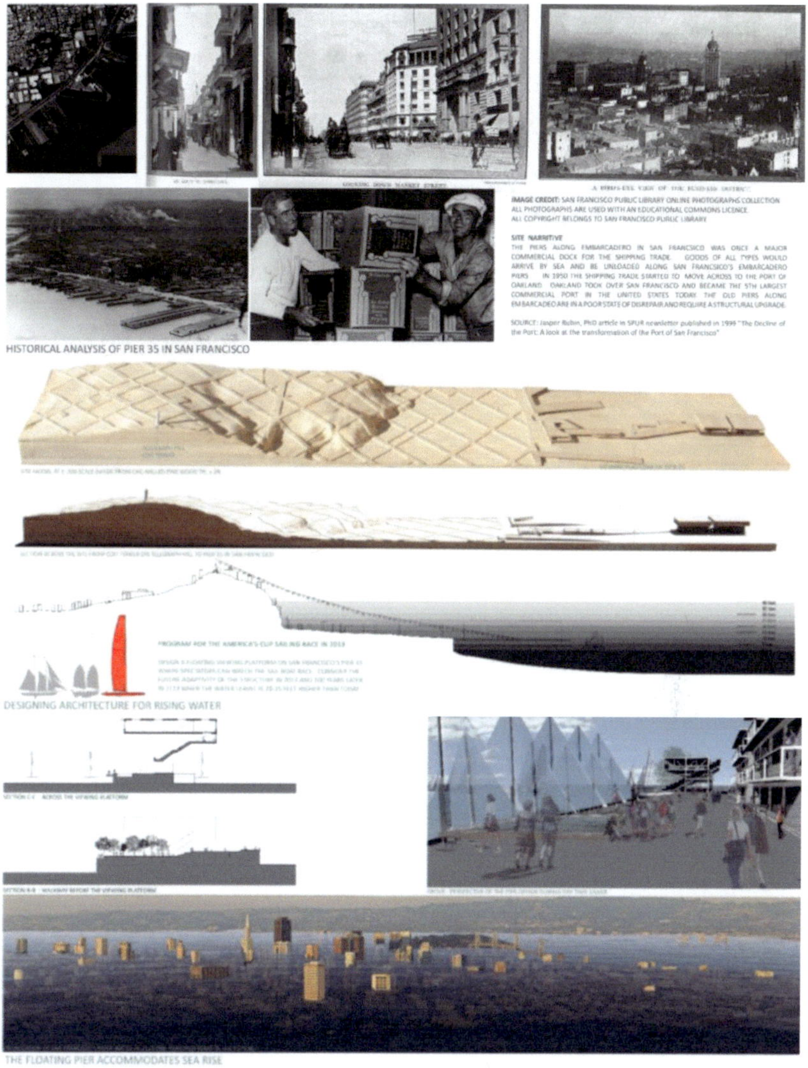

Figure 26: Watermarks part II.

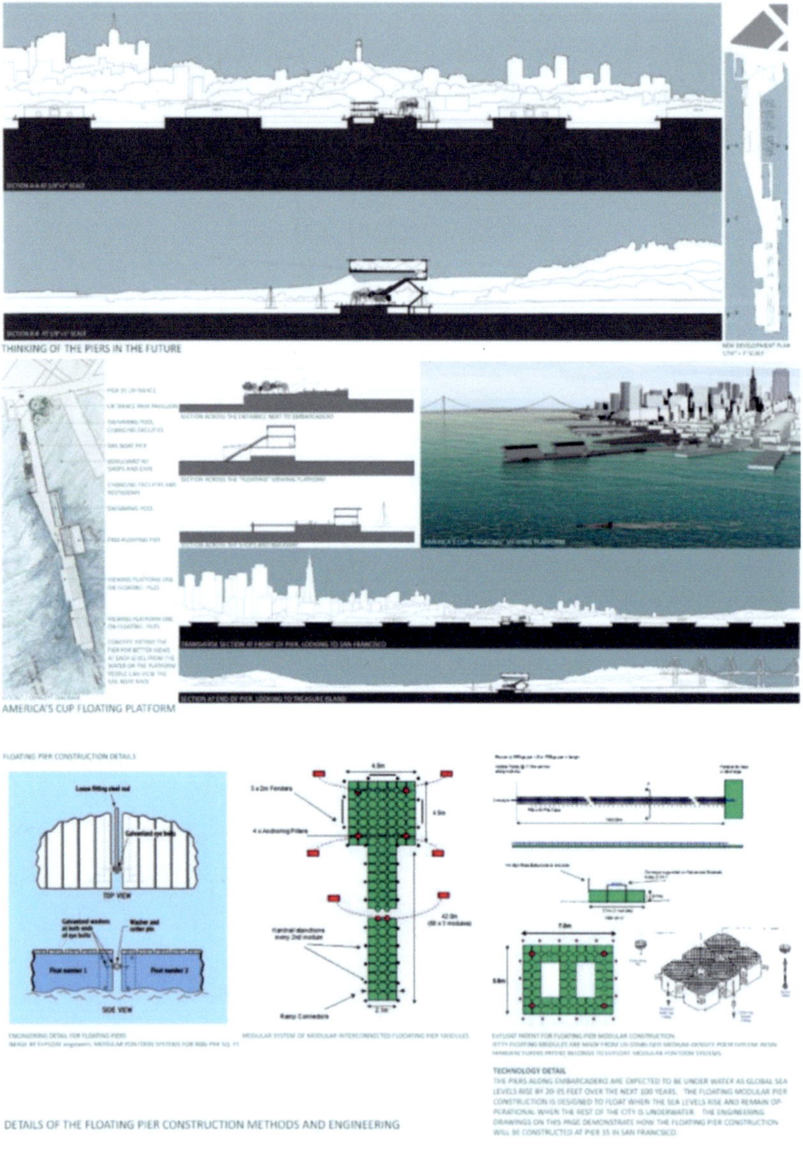

Figure 27: Watermarks part III.

Media

[1] "Why Architects Visit Rome in the 21st Century" in the "GenslerOn Cities" professional online journal for design professionals. Link: http://www.gensleron.com/cities/2016/1/25/why-architects-visit-rome-in-the-21stcentury.html . This article builds on Mark's December 2015 Bibliotheca Hertziana conference presentation in Rome about why Rome still matters to contemporary architects.

[2] Mark Andrew Kelly's blog, is where you can find all of his timelapse videos of how he has been spending his time in Rome: http://www.markkellyarchitect.com/blog/

[3] Meet the Artists- Interview https://britishschoolatrome.wordpress.com/tag/mark-andrew-kelly/

[4] Using Lost wax in the 21st century https://britishschoolatrome.wordpress.com/2015/12/03/using-the-lost-wax-method-ofbronze-casting-in-the-21st-century/

[5] Conference at Bibliotheca Hertziana http://www.bsr.ac.uk/mark-kelly-at-the-bibliotheca-hertziana

[6] UC Berkeley Alumni newsletter http://ced.berkeley.edu/newsletters/e-news/12-9-2015__

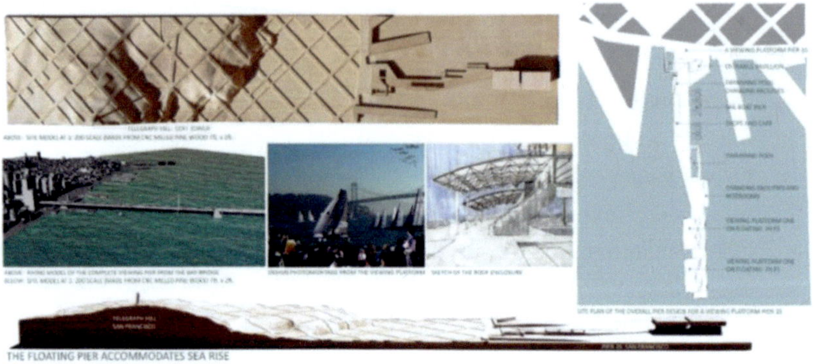

Figure 28: Watermarks part IV.

Migration in Waterscapes: From Royal Model to New-Scapes in Copenhagen

Marina Mihaila [1], Cristian Banica [2]

[1] *University of Architecture and Urbanism "Ion Mincu" Bucharest (Romania).*
[2] *Center for Studies in Contemporary Architecture, Bucharest (Romania).*

Abstract

Presentation focuses on Migration in waterscapes: from royal model to new-scapes in Copenhagen as working paper, applying observations of previous studies of the authors on architectural cultural model of Copenhagen, but also discussing on a successive successful local policy of spatial development.

For the 21st century architecture, Copenhagen is a perfect urban place morphed into a multitude of successive identity landscapes, writing a new brand topographical architectural front on a substantial heritage background and patrimonial structure of a royal city presence. In the last almost 10 years Copenhagen has encouraged the building of a brand new architectural presence; a multitude of buildings have appeared and a series of new regional and local architectural firms expressed and experimented new shapes, functions and in-out public spaces.

The paper/presentation refers to the topographical research on the main Copenhagen channel, containing both patrimonial places from Royal Plaza and Royal Library to the presence of the fortress, fortifications and industrial halls.

Method of research is based on observations made on the Copenhagen channel in 2006 and 2015 (study trips made by the authors), correlated with an architectural investigation (own method and approach of the authors) on architectural insertions on the main Copenhagen channel. Also, presentation will focus on Copenhagen policy of writing urban development and urban space, since the 50s with the Finger Plan to the present with the Sustainia Project for 2025 but also discussing the importance of architecture policy declaration in Copenhagen City of Architecture (The Architecture Policy of The City of Copenhagen).

Data of presentation research conclusions in the last ten years (2006-2015):

- Areal of city representation: urban-architectural presence to the main Copenhagen sea channel.
- Waterscapes: rewriting the water front => successive landscape image to the main sea channel
- Migration: defined as an urban-architectural densification and master planning local cultural policy intention in the built water areal.
- New developed living models: towards zero energy city, cycling and walking city, fashion-architectural brand city, architectural patrimonial-royal model as public space status encouraging new 21st century architectures.

Keywords: Architecture, architecture policy, sustainability, waterscapes, urban innovation, new-scapes, Copenhagen.

Vita

MA Architect Cristian Bănică is director and partner architect within office for architecture, urban planning and project management Arhitectonik2000 Bucharest, where he authored more than 70 notable projects as architect and project manager. He holds MA Architect title from University of Architecture and Urbanism "Ion Mincu" Bucharest, and he is graduated MSc PM in Architecture from Heriot Watt University and Edinburgh University. He is full chartered member of Romanian Order of Architects (OAR), and chartered member of Registered Urban Planner of Romania (RUR). He authored and co-authored few important papers in the field of architecture, sustainability and PM.

PhD Architect Marina Mihăilă is lecturer in Synthesis of Architectural Design Studio (Master) within Faculty of Architecture – University of Architecture and Urbanism "Ion Mincu" Bucharest, and member of the university academic staff since 2001. Marina Mihăilă is also a practitioner architect – partner within office for architecture, urban planning and PM Arhitectonik2000 Bucharest, and she holds the position of research director within Association Center for Studies in Contemporary Architecture Bucharest. She is full chartered member and founder of Romanian Order of Architects (OAR), and chartered member of Registered Urban Planner of Romania (RUR). Marina Mihăilă authored 2 books and few important international papers, holds a PhD title (2011, UAUIM) and she directed postdoctoral project Types of innovation in cultural spaces.]working with/in[cultural spaces_tradition and innovation. (2013-2015, project number PN-II-RU-PD-2012-3 -0515, Romanian Ministry of Education – CNCS-UEFISCDI); her research topics are: studies regarding architectural density and cultural space, anthropological architecture, city values and contemporary theory of architecture. She authored more than 50 notable built projects architectural and urban designs (among which: medium and high-rise collective housing, mixed functioned buildings, industrial mixed parks, conference halls, office buildings, National architectural flagship design and built projects for 4 companies, etc.). Her status is "architect as artist" as professional belief, also the name of her topic page: https://architectasartist.wordpress.com/ A description of her research work here:
https://cultureandspace.wordpress.com/2015/05/27/overview-research-continuity/

Figure 29: Copenhagen, architectural landscape of city main channel (panoramic views) by Marina Mihăilă & Cristian Bănică, 2015.

References

[1] Mihăilă M. (2012-2015). Types of innovation in cultural spaces.]working with/in[cultural spaces_tradition and innovation. (2013-2015, Postdoctoral Project Research, project number PN-II-RU-PD-2012-3 -0515, Romanian Ministry of Education – CNCS-UEFISCDI). https://cultureandspace.wordpress.com/

[2] Mihăilă M. (2015). Water-scapes in Architectural Design Thinking – a Discussion on Water as Conceptual Cultural Sustainable Design Element, 2015/12 SMC Sustainable Mediterranean Construction Magazine no.2, Luciano Editore, Naples, Italy. ISSN on-line 2420-8213, pp. 009-013.

[3] http://www.sustainablemediterraneanconstruction.eu/SMC/The_Magazine_n.2_files/Smc%20N.2_pap_01.pdf

[4] Mihăilă M., Bănică C. (2015). Sustainability: notions, directions, dictionaries and rewriting application terms in architecture, pp.36-46, in Proceedings Book – Conference: Conference URBAN INCERC INCD: 10th edition of the research conference on constructions, economy of buildings, architecture, urban and territorial development | INNOVATIVE CONCEPTS AND SOLUTIONS FOR ADAPTING TO CLIMATE CHANGES, BUCHAREST, OCTOBER 2015., At URBAN INCERC INCD BUCHAREST, ROMANIA, Editura INCD URBAN-INCERC, OCTOBER 2015, ISSN 2393-3208.

[5] https://www.researchgate.net/publication/283636753_Sustainability_notions_directions_dictionaries_and_rewriting_application_terms_in_architecture?ev=prf_pub

[6] Bănică C., Mihăilă M. (2016). Setup Propositions in Architectural Design Thinking: Towards Control and Project Management, in volume NASHS2015: New Approaches in Social and Humanistic Sciences, EDITOGRAFICA Italy (former MEDIMOND Monduzzi Editore International Proceedings), May 2016, with ISBN: 9788875877286

[7] https://www.researchgate.net/publication/291829764_Setup_Propositions_in_Architectural_Design_Thinking_Towards_Control_and_Project_Management

[8] Bănică C. (2011). Urban transformers _ A case study on urban transformation. Bucharest between two worlds and the forces shaping the city / Urban Transformers_ Studiu de caz transformare urbană. Bucureştiul între două lumi şi forţele ce transformă oraşul, pp.93-110, ARGUMENT, Editura Universitară „Ion Mincu" – Bucureşti, Ion Mincu University Publishing House, ISSN: 2067 - 4252, http://argument.uauim.ro/f/argument/attachment/ARG3_Banica.pdf
Gnielka, S. (1999). Titel der Veröffentlichung. Mannheim.

The Weather Radar Approach for Rain Precipitation Estimation in Urban Areas

Mario Montopoli

National Research Council of Italy (CNR), Institute of Atmospheric Sciences and Climate (ISAC)

Abstract

Real-time monitoring and quantitative nowcasting of severe precipitation events is nowadays an essential duty for meteorological services operating in operational contexts. Ground based systems, comprising imaging weather radars and punctual sources (e.g. rain gauges, disdrometers and vertical profilers) can be used in a complementary way to have a complete description of the rain precipitating at the ground or nearby. However, among other sources of radar uncertainties (i.e. path attenuation, calibration issues, partial beam blocking, radio interference, particle size distribution variability, ground clutter) a gap exists between measurements performed by weather radars at some altitude from the ground and those produced by the punctual instruments in proximity of the ground. This gap is even more exacerbated in those situations of complex orography where the pointing of the radar antenna needs to be increased to avoid clutter contaminations (i.e. unwanted returns from the ground) in the acquired data while maintaining a reasonable spatial coverage.

In the presentation held, given the multisciplinarity of the attendees, It has been given the basic principles of the main tools nowasays used to monitor the atmospheric rain precipitation with a special enphasis to weather radars. Some applicaations have been shown especially those in the Roman area where the ISAC-CNR operated with its own radar system (see figure below).

Vita

http://www.isac.cnr.it/en/users/mario-montopoli

My back ground is in general on information theory; remote sensing of the atmosphere; inverse problems; weather radar processing algorithms and interpretation of its measured quantities; electromagnetic radiative transfer and scattering models from oblate particles (e.g. hydrometeors); passive active satellite observations of precipitation; techniques for validation studies of environmental quantities (e.g. dual or triple colocation validation schemes); statistical analysis. In addition, I support the teaching in the introductory university courses of antennas, radio-propagation and remote sensing.

In my past experience I had the opportunity to work in several EU funded projects handling large geo-science datasets dealing with: GPS (Global Position System) studies to retrieve water vapour profile to support SAR (synthetic Aperture Radar) measurements (ESA-METAWAVE project); electromagnetic Lunar radiative transfer models of its stratigraphy to support future microwave passive measurements of the Lunar soil (ESA-ESMO project); Joint use of models, multi-platform observations and ad-hoc developed algorithms, for the quantitative estimates of ash features during on going volcanic Eruptions (FP7-EU RASHCAST project);

Dr. Montopoli has authored 14 peer-to-peer papers, co-authored 23 peer-to-peer papers published on international journals, submitted 66 conference contributions and 2 book chapters. In 2010, Dr. Montopoli won the price for the best submitted paper on international journal within the European Radar conference (ERAD).

Figure 30: Slide from the presentation.

Smartness, Technologies and Natural Risks

Beniamino Murgante

University of Basilicata, Potenza

Abstract

https://de.scribd.com/doc/315613231/Smartness-technologies-and-natural-risks

Very often the concept of smart city is strongly related to the flourishing of mobile applications, stressing the technological aspects and a top-down approach of high-tech centralized control systems capable of resolving all the urban issues, completely forgetting the essence of a city with its connected problems. The real challenge in future years will be a huge increase in the urban population and the changes this will produce in energy and resource consumption. It is fundamental to manage this phenomenon with clever approaches in order to guarantee a better management of resources and their sustainable access to present and future generations. This chapter develops some considerations on these aspects, trying to insert the technological issues within a framework closer to planning and with attention to the social impact.

Vita

Beniamino Murgante presently is professor of spatial panning School of Engineering of the University of Basilicata.

He took his PhD in "Sciences and methods for European cities and territory" at the Department of Civil engineer of the University of Pisa and he carried out other researches in Lyon at the Laboratory for Information System Engineering directed by Robert Laurini.

He is the author of more than 100 international publications in the field of technology applied to the city and the territory.

Member of the Editorial Board of several international journals, of scientific committees of many national and international conferences and of the scientific council of some national and international organizations.

Co-program committee Chair of the International Conference on Computational Science and Its Applications (ICCSA).

More information is available on the personal web page:

http://oldwww.unibas.it/utenti/murgante/Benny.html.

TRA_MED – Mediterranean Transhumances Immigrants shepherds in Mediterranean pastoralism

Michele Nori

Migration Policy Centre - European University Institute, Firenze

Abstract

In third millennium Europe, agricultural activities are increasingly carried out by foreign workers, who contribute their skills and experience to a sector that has been showing limited attractiveness to new European generations – despite the fact that food production and land management are two basic societal functions. Such phenomenon is visible throughout the European Union - from milk supply chains, with significant presence of Indian and Pakistani immigrants, to picking activities undertaken by African seasonal workers to other sectors with other groups more or less specialized in specific activities. Pastoralism represents a specific case of this phenomenon, with a significant and growing presence of immigrants on European pasturelands, especially during the period of seasonal transhumance. Such dynamic is common to southern European countries - Italy, but also Spain, Greece, France and Portugal – where agricultural represent a main pillar of the socio-cultural and economic identities, and pastoralism is a longstanding and traditional activity in the Mediterranean setting. The recent reform of the Common Agricultural Policy (CAP) - which involves about 40% of the total EU budget – allocates an increasing emphasis on the proper management of mountainous areas, agro-ecologically fragile territories and their biodiversity –often inhabited and operated by pastoralists. These territories provide as well important opportunities for the tourism market which benefits from local natural resources. In addition, the typical quality meat as well as dairy products provided by pastoral systems, often with organic and geographic certification, are on increasing demand by consumers. Through its socio-ecological services pastoralism contributes importantly to maintaining marginal territories alive and productive. While we face thus a growing societal demand for the services as well as the products of pastoralism, this is not reflected by improving living and working conditions of shepherds in neither economic, nor social terms. Current dynamics rather indicate that local rural youth often tend to opt out from the sector and look for urban-based livelihoods and professions - leaving thus pasturelands exposed to a lack of generational renewal. In such context an increasing presence of immigrant shepherds is reported. The shepherds who come to provide their workforce on southern European pastures often originate from pastoral communities in other parts of the Mediterranean region, Morocco, Albania and the Balkans, but also from other countries such as Romania, Bulgaria and Moldova. The living and working conditions of these immigrant shepherds are often extremely difficult and precarious, but cases of virtuous contracts and dignified conditions are also reported. The research initiative 'Mediterranean Transhumances' is concerned with assessing ongoing dynamics affecting pastoralism in the Mediterranean, so to provide a more effective understanding of the opportunities, the risks and the difficulties related to the increasing presence of immigrants in this domain. The aim is to contribute to the development of appropriate policies at local as well as at European level.

Vita

http://www.migrationpolicycentre.eu/people/michele-nori/

Michele Nori is a Tropical Agronomist (University of Florence – IT) with a further specialization in Rural Sociology (PhD at Wageningen University – NL).

With about 20 years of field experience on the technical as well as socio-economic aspects of natural resource management in different regions of the world (mostly Africa, but also Asia and Latin America), where he has mainly dealt with the livelihood systems of pastoral communities.

By intensively working at the interface between academic, field as decision-making levels, he has developed an 'horizontal career', based on collaborations with and consultancies to a number of organizations at different levels: civil society, UN agencies, research institutes, donors offices as well as the European Commission Development Cooperation and EU Delegations.

His list of publications ranges from scientific papers to technical notes to advocacy documents on matters relates to sustainable pastoralism.

His current concern is to adequately inform policy decision-making on aspects of rural development, food security and natural resource management in the Mediterranean region – a domain where migrations plays an increasingly important role.

He has been awarded a Marie Curie fellowship to investigate on the increasingly relevant role of immigrant shepherds for the sustainability of pastoral systems in Mediterranean EU countries – a specific case study within the debate about migrants' contributions to farming and rural development in Europe.

Villa Gamberaia and the Parterre D'eau of Princess Ghyka (1896-1925): Water as Heritage in a Potentially Hazardous Environment

Patricia Osmond

Iowa State University

Abstract

This paper focuses on the iconic parterre d'eau at the Villa Gamberaia (Florence), designed by the Romanian Princess Catherine Jeanne Ghyka in the years following her purchase of the villa in 1896. Here, the pools of water, placidly mirroring the sky and the surrounding plants and shrubs, recall the long tradition of otium in villa life and our own notions of leisure architecture. This special feature of the garden, however, will also be considered within a broader context, that is, the hazards associated with the creation and maintenance of an elaborate hydraulic system supplying water to the villa and gardens, and, as seen in the iconography of the Neptune grotto, with memories of water's potentially destructive powers.

Vita

Dr. Patricia Osmond (Rome, Italy / Cambridge, MA, USA) is an historian of the Renaissance and Early Modern Europe. Her research and publications focus chiefly on aspects of the classical tradition and Roman historiography, but she has also written on, and promoted studies of, the Villa Gamberaia

(Florence), owned by the Romanian Princess Catherine Jeanne Ghyka from 1896 to 1925. She has edited two publications on the villa—Villa Gamberaia: Sources and Interpretations, vol. 22 (2002) of Studies in the History of Gardens & Designed Landscapes, and Revisiting the Gamberaia: An Anthology of Essays (Florence 2004; repr. 2014) —and contributed to "Catena," the Digital Archive of Historic Gardens and Landscapes (Bard Graduate Center), http://catena.bgc.bard.edu (2005). She is currently Affiliate Assoc. Professor, Dept. of Art and Visual Culture, College of Design, Iowa State University, and former Resident Director of the ISU Rome Program.

Virtual Museum of the Tiber Valley

Eva Pietroni

CNR ITABC

Abstract

The Virtual Museum of the Tiber Valley, developed by the National Research Council - Institute of Technologies applied to Cultural Heritage (CNR ITABC), is an integrated system of knowledge, promotion and communication of the cultural landscape of the Tiber middle valley, in the northern area of Rome, between Monte Soratte and Fidene and between cities of Sacrofano and Palombara Sabina (about 60 x 40 km). Starting from the existing data collection, the digital documentation and the topographic survey, 3D reconstructions have been realized at different scale: - the landscape, in its geological and historical evolution (since 3 million years ago to the present); - some archaeological sites (Lucus Feroniae, Villa Volusii); - areas of historical and natural interest (Soratte Mountain and the Tiber-Farfa Natural Reserve). The Tiber emerges from various points of view: the geo-morphological environment, the ecosystem, the settlement and human activities, the mythology linked to the river, the rituals and iconography, the literary, natural and archaeological landscape. Storytelling, evocation and interaction modes involve the visitor in an emotional and multi-sensorial experience, through a language that combines science, technology and art. Various multimedia applications have been developed and put permanently in museums, both of the city of Rome and of the peripheral territory:

- The Museo of River, Nazzano
- The new archaeological museum of Lucus Feroniae, Capena
- The library of "Società Geografica Italiana" at Villa Celimontana, Rome.
- The National Etruscan Museum of Villa Giulia, Rome.

A spectacular virtual reality installation with gesture-based interaction has been created for the Museum of Villa Giulia, conceived as attractive point, to convey the public's interest in visiting the actual sites and raise their affection towards our cultural heritage.

Vita

Eva Pietroni is an art historian, a conservation specialist of Cultural Heritage and a musician.

Since 1998 she has been developing research activity in the Institute of Sciences and Technologies of Cognition and in the Institute of Technologies Applied to Cultural Heritage of CNR. She is now a researcher, responsible and co-director of the Virtual Heritage Lab of CNR ITABC. The ambits of research and application are referred to the study, documentation, communication and valorization of Cultural Heritage, using the digital interactive technologies, with particular attention to museum and virtual environments. In the field of virtual reality she works in projects for the virtual reconstructions of archaeological landscapes, urban and monumental sites, artistic contexts. Different types of data, acquired directly on the field are elaborated and integrated in VR environments. Particular interest is oriented in the development of complex behaviours, multiuser domains, natural interaction and artificial life in VR ecosystems.

She is also interested in cultural applications for portable technologies (smartphone and tablet) dedicated to arts, sites and museums. She has collaborated, (since 2008-2009 for coordination and technical direction rule), to some VR applications that have obtained important international appreciations: Casa dei Vettii in Pompeii, the archaeological landscape of Aksum in Ethiopia (2001), Giotto's Scrovegni Chapel in Padua (2002-2003, Best paper award for cultural merit at the eighth international conference VSMM, virtual systems and multimedia, 2002), the Appia Antica project (2003-2005), the reconstruction of archaeological landscape of Tell Mozan in Syria (2004), the Multiuser domain VR project of Via Flaminia Antica (2006-2008, 1° E-Content Award in December 2008), the "Integrated Technologies of Robotics and Virtual environment in Archaeology project" (2006-2009, Best Paper Award VSMM 2008), Matera: tales of a city project (2009-2012), "The Rule Confirmation": virtual experience among the characters of Giotto's work (2010),Etruscanning 3D (European project, 2011-2013), The Virtual Museum of Tiber valley (2011-2013), Virtual Museum Transnational network (V-Must), 2011-2014.

Main technical skills: project coordination, 2d and 3D computer graphics, data optimization for real time graphic engines, laser scanner acquisition and data elaboration, photogrammetry and photomodelling, editing video.

Urban Pastoralism – Spontaneous Large Scale Phenomenon

Roxana Triboi

"Ion Mincu" University of Architecture and Urbanism

Abstract

The Eastern European context has an important potential for developing this form of sustainable urbanism. In the last years, due to a combination of factors like chaotic fragmentation of the periphery, development of urban gaps or waste land, the abandonment of agricultural exploitation of arable land, the pastures developed and prospered mostly in the green pockets of urban peripheries and therefore the urban pastoralism represents the most important expression of city-agriculture interaction in this geographical region.

Rather than seize the potential of this practice endangered and use it as a tool for sustainable urban management, local authorities tolerate at best regulate this practice without regulation vision. In the absence of coherent management, this practice exist mostly in its spontaneous form, fact that generates some negative consequences like: the persistence of conflict of interest issues, negative ecological effects caused by spontaneous grazing, negative perception of the urban population on this phenomenon and finally his extinction.

The possible benefits of this spontaneous practice in the peripheral urban areas are multiple like: establishing biological corridors, developing biodiversity in general, articulating unused green lots, providing a picturesque landscape, preventing reforestation, supporting short supply circuits (milk, cheese, meat), ensuring the social dimension of green space and providing a pedagogical function on agriculture and food. The recognition of cultural and patrimonial value of this practice, the use of "construction designated land" and private and public properties as pasture areas by urban planners and policy makers by creating new types of zoning that includes such forms of temporary or permanent land use, the identification of the resulted and maintenance management of ecological corridors could transform Urban Pastoralism in an important tool for sustainable urbanism and for creating unity in a very fragmented and heterogenic landscape, because it values both the agricultural, social, economical and ecological dimension of urban lands.

Vita

Roxana Maria Triboi is an urban planner with more than 10 years experience in urban planning and landscape architecture in Romania and France, with a special interest in food planning and environmental issues.

Currently a PhD student at "Ion Mincu" University of Architecture and Urbanism of Bucharest, she is researching on the specificity of urban agriculture in Romania and Eastern Europe with a focus on urban pastoralism, as a possible environmental tool for a sustainable urban planning.

Figure 31: Urban pastoralism images.